Lecture Notes in Networks and Systems

Volume 327

Series Editor

Janusz Kacprzyk, Systems Research Institute, Polish Academy of Sciences, Warsaw, Poland

Advisory Editors

Fernando Gomide, Department of Computer Engineering and Automation—DCA, School of Electrical and Computer Engineering—FEEC, University of Campinas—UNICAMP, São Paulo, Brazil

Okyay Kaynak, Department of Electrical and Electronic Engineering, Bogazici University, Istanbul, Turkey

Derong Liu, Department of Electrical and Computer Engineering, University of Illinois at Chicago, Chicago, USA; Institute of Automation, Chinese Academy of Sciences, Beijing, China

Witold Pedrycz, Department of Electrical and Computer Engineering, University of Alberta, Alberta, Canada; Systems Research Institute, Polish Academy of Sciences, Warsaw, Poland

Marios M. Polycarpou, Department of Electrical and Computer Engineering, KIOS Research Center for Intelligent Systems and Networks, University of Cyprus, Nicosia, Cyprus

Imre J. Rudas, Óbuda University, Budapest, Hungary

Jun Wang, Department of Computer Science, City University of Hong Kong, Kowloon, Hong Kong

AF864049

The series "Lecture Notes in Networks and Systems" publishes the latest developments in Networks and Systems—quickly, informally and with high quality. Original research reported in proceedings and post-proceedings represents the core of LNNS.

Volumes published in LNNS embrace all aspects and subfields of, as well as new challenges in, Networks and Systems.

The series contains proceedings and edited volumes in systems and networks, spanning the areas of Cyber-Physical Systems, Autonomous Systems, Sensor Networks, Control Systems, Energy Systems, Automotive Systems, Biological Systems, Vehicular Networking and Connected Vehicles, Aerospace Systems, Automation, Manufacturing, Smart Grids, Nonlinear Systems, Power Systems, Robotics, Social Systems, Economic Systems and other. Of particular value to both the contributors and the readership are the short publication timeframe and the world-wide distribution and exposure which enable both a wide and rapid dissemination of research output.

The series covers the theory, applications, and perspectives on the state of the art and future developments relevant to systems and networks, decision making, control, complex processes and related areas, as embedded in the fields of interdisciplinary and applied sciences, engineering, computer science, physics, economics, social, and life sciences, as well as the paradigms and methodologies behind them.

Indexed by SCOPUS, INSPEC, WTI Frankfurt eG, zbMATH, SCImago.

All books published in the series are submitted for consideration in Web of Science.

More information about this series at http://www.springer.com/series/15179

Kenji Matsui · Sigeru Omatu ·
Tan Yigitcanlar · Sara Rodríguez González
Editors

Distributed Computing and Artificial Intelligence, Volume 1: 18th International Conference

Springer

Editors
Kenji Matsui
Faculty of Robotics and Design
Osaka Institute of Technology
Osaka, Japan

Sigeru Omatu
Graduate School
Hiroshima University
Higashi-Hiroshima, Osaka, Japan

Tan Yigitcanlar
School of Architecture
and Built Environment
Queensland University of Technology
Brisbane, Australia

Sara Rodríguez González
BISITE, Digital Innovation Hub
University of Salamanca
Salamanca, Spain

ISSN 2367-3370 ISSN 2367-3389 (electronic)
Lecture Notes in Networks and Systems
ISBN 978-3-030-86260-2 ISBN 978-3-030-86261-9 (eBook)
https://doi.org/10.1007/978-3-030-86261-9

© The Editor(s) (if applicable) and The Author(s), under exclusive license
to Springer Nature Switzerland AG 2022
This work is subject to copyright. All rights are solely and exclusively licensed by the Publisher, whether the whole or part of the material is concerned, specifically the rights of translation, reprinting, reuse of illustrations, recitation, broadcasting, reproduction on microfilms or in any other physical way, and transmission or information storage and retrieval, electronic adaptation, computer software, or by similar or dissimilar methodology now known or hereafter developed.
The use of general descriptive names, registered names, trademarks, service marks, etc. in this publication does not imply, even in the absence of a specific statement, that such names are exempt from the relevant protective laws and regulations and therefore free for general use.
The publisher, the authors and the editors are safe to assume that the advice and information in this book are believed to be true and accurate at the date of publication. Neither the publisher nor the authors or the editors give a warranty, expressed or implied, with respect to the material contained herein or for any errors or omissions that may have been made. The publisher remains neutral with regard to jurisdictional claims in published maps and institutional affiliations.

This Springer imprint is published by the registered company Springer Nature Switzerland AG
The registered company address is: Gewerbestrasse 11, 6330 Cham, Switzerland

Preface

The 18th International Symposium on Distributed Computing and Artificial Intelligence 2021 (DCAI 2021) is a forum to exchange of ideas between scientists and technicians from both academic and business areas is essential to facilitate the development of systems that meet the demands of today's society. The technology transfer in fields such as distributed computing or artificial intelligence is still a challenge, and for that reason, this type of contributions will be specially considered in this symposium. Distributed computing performs an increasingly important role in modern signal/data processing, information fusion, and electronics engineering (e.g., electronic commerce, mobile communications, and wireless devices). Particularly, applying artificial intelligence in distributed environments is becoming an element of high added value and economic potential.

Research on intelligent distributed systems has matured during the last decade and many effective applications are now deployed. Nowadays, technologies, such as Internet of things (IoT), Industrial Internet of things (IIoT), big data, blockchain, distributed computing in general, are changing constantly because of the large research and technical effort being undertaken in both universities and businesses. Most computing systems from personal laptops to edge/fog/cloud computing systems are available for parallel and distributed computing. This conference is the forum in which to present application of innovative techniques to complex problems in all these fields.

This year's technical program will present both high quality and diversity, with contributions in well-established and evolving areas of research. Specifically, 55 papers were submitted to main track and special sessions, by authors from 24 different countries (Angola, Brazil, Bulgaria, Colombia, Czechia, Denmark, Ecuador, France, Germany, Greece, India, Italy, Japan, Latvia, Lebanon, Mexico, Poland, Portugal, Russia, Spain, Sweden, Tunisia, Turkey, USA), representing a truly "wide area network" of research activity. The DCAI'21 technical program has selected 21 papers and, as in past editions, it will be special issues in ranked journals such as electronics, sensors, systems, robotics, mathematical biosciences, and ADCAIJ. These special issues will cover extended versions of the most highly regarded works. Moreover, DCAI'21 Special Sessions have been a very useful tool to complement the regular program with new or emerging topics of particular interest to the participating community.

We would like to thank all the contributing authors, the members of the program committee, the sponsors (IBM, Armundia Group, EurAI, AEPIA, APPIA, CINI, OIT, UGR, HU, SCU, USAL, AIR Institute, and UNIVAQ), the organizing committee of the University of Salamanca for their hard and highly valuable work, the funding supporting of the project "Intelligent and sustainable mobility supported by multi-agent systems and edge computing (InEDGEMobility): Toward Sustainable Intelligent Mobility: Blockchain-based framework for IoT Security," Reference: RTI2018-095390-B-C32, financed by the Spanish Ministry of Science, Innovation and Universities (MCIU), the State Research Agency (AEI) and the European Regional Development Fund (FEDER), and finally, the local organization members, and the program committee members for their hard work, which was essential for the success of DCAI'21.

October 2021

Kenji Matsui
Sigeru Omatu
Tan Yigitcanlar
Sara Rodríguez

Organization

Honorary Chairman

Masataka Inoue	President of Osaka Institute of Technology, Japan

Advisory Board

Yuncheng Dong	Sichuan University, China
Francisco Herrera	University of Granada, Spain
Enrique Herrera Viedma	University of Granada, Spain
Kenji Matsui	Osaka Institute of Technology, Japan
Sigeru Omatu	Hiroshima University, Japan

Program Committee Chairs

Tiancheng Li	Northwestern Polytechnical University, China
Tan Yigitcanlar	Queensland University of Technology, Australia

Organizing Committee Chair

Sara Rodríguez	University of Salamanca, Spain

Workshop Chair

José Manuel Machado	University of Minho, Portugal

Program Committee

Ana Almeida	ISEP-IPP, Portugal
Gustavo Almeida	Instituto Federal do Espírito Santo, Brazil
Ricardo Alonso	University of Salamanca, Spain

Giner Alor Hernandez	Instituto Tecnologico de Orizaba, Mexico
Cesar Analide	University of Minho, Portugal
Luis Antunes	GUESS/LabMAg/Univ. Lisboa, Portugal
Fidel Aznar	Universidad de Alicante, Spain
Zbigniew Banaszak	Warsaw University of Technology, Faculty of Management, Dept. of Business Informatics, Poland
Olfa Belkahla Driss	University of Manouba, Tunisia
Carmen Benavides	University of León, Spain
Holger Billhardt	Universidad Rey Juan Carlos, Spain
Amel Borgi	ISI/LIPAH, Université de Tunis El Manar, Tunisia
Pierre Borne	Ecole Centrale de Lille, France
Lourdes Borrajo	University of Vigo, Spain
Adel Boukhadra	National High School of Computer Science, Algeria
Edgardo Bucciarelli	University of Chieti-Pescara, Italy
Juan Carlos Burguillo	University of Vigo, Spain
Francisco Javier Calle	Departamento de Informática. Universidad Carlos III de Madrid, Spain
Rui Camacho	University of Porto, Portugal
Juana Canul Reich	Universidad Juarez Autonoma de Tabasco, México
Wen Cao	Chang'an University, China
Davide Carneiro	University of Minho, Portugal
Carlos Carrascosa	GTI-IA DSIC Universidad Politecnica de Valencia, Spain
Roberto Casado Vara	University of Salamanca, Spain
Luis Castillo	Autonomous University of Manizales, Colombia
Camelia Chira	Babes-Bolyai University, Romania
Rafael Corchuelo	University of Seville, Spain
Paulo Cortez	University of Minho, Portugal
Ângelo Costa	University of Minho, Portugal
Stefania Costantini	Dipartimento di Ingegneria e Scienze dell'Informazione e Matematica, Univ. dell'Aquila, Italy
Kai Da	National University of Defense Technology, China
Giovanni De Gasperis	Dipartimento di Ingegneria e Scienze dell'Informazione e Matematica, Italy
Fernando De La Prieta	University of Salamanca, Spain
Carlos Alejandro De Luna-Ortega	Universidad Politecnica de Aguascalientes, Mexico
Raffaele Dell'Aversana	Università "D'Annunzio" di Chieti-Pescara, Italy
Richard Demo Souza	Federal University of Santa Catrina, Brazil

Kapal Dev	Cork Institute of Technology, Ireland
Fernando Diaz	University of Valladolid, Spain
Worawan Diaz Carballo	Thammasat University, Thailand
Youcef Djenouri	LRIA_USTHB, Algeria
António Jorge Do Nascimento Morais	Universidade Aberta, Portugal
Ramon Fabregat	Universitat de Girona, Spain
Jiande Fan	Shenzhen University, China
Ana Faria	ISEP, Portugal
Pedro Faria	Polytechnic of Porto, Portugal
Florentino Fdez-Riverola	University of Vigo, Spain
Alberto Fernandez	CETINIA. University Rey Juan Carlos, Spain
Peter Forbrig	University of Rostock, Germany
Toru Fujinaka	Hiroshima University, Japan
Svitlana Galeshchuk	Nova Southeastern University, USA
Jesús García	University Carlos III Madrid, Spain
Francisco Garcia-Sanchez	University of Murcia, Spain
Marisol García Valls	Universitat Politècnica de València, Spain
Irina Georgescu	Academy of Economic Studies, Romania
Abdallah Ghourabi	Higher School of Telecommunications SupCom, Tunisia
Ana Belén Gil González	University of Salamanca, Spain
Arkadiusz Gola	Lublin University of Technology, Poland
Juan Gomez Romero	University of Granada, Spain
Evelio Gonzalez	Universidad de La Laguna, Spain
Angélica González Arrieta	Universidad de Salamanca, Spain
Alfonso González Briones	Universidad de Salamanca, Spain
Carina Gonzalez Gonzalez	Universidad de La Laguna, Spain
David Griol	Universidad Carlos III de Madrid, Spain
Zhaoxia Guo	Sichuan University, China
Elena Hernández Nieves	Universidad de Salamanca, Spain
Felipe Hernández Perlines	Universidad de Castilla-La Mancha, Spain
Aurélie Hurault	IRIT - ENSEEIHT, France
Elisa Huzita	State University of Maringa, Brazil
Gustavo Isaza	University of Caldas, Colombia
Patricia Jiménez	Universidad de Huelva, Spain
Bo Nørregaard Jørgensen	University of Southern Denmark, Denmark
Günter Koch	Humboldt Cosmos Multiversity & GRASPnetwork, Karlsruhe University, Faculty of Informatics, Germany
Bo Noerregaard Joergensen	University of Southern Denmark, Denmark
Vicente Julian	Universitat Politècnica de València, Spain
Geylani Kardas	Ege University International Computer Institute, Turkey
Amin Khan	UiT the Arctic University of Norway, Norway

Naoufel Khayati	COSMOS Laboratory - ENSI, Tunisia
Egons Lavendelis	Riga Technical University, Latvian
Rosalia Laza	Universidad de Vigo, Spain
Tiancheng Li	Northwestern Polytechnical University, China
Weifeng Liu	Shaanxi University of Science and Technology, China
Ivan Lopez-Arevalo	Cinvestav - Tamaulipas, Mexico
Daniel López-Sánchez	BISITE, Spain
Ramdane Maamri	LIRE laboratory UC Constantine2- Abdelhamid Mehri, Algeria
Benedita Malheiro	Instituto Superior de Engenharia do Porto, Portugal
Eleni Mangina	UCD, Ireland
Fabio Marques	University of Aveiro, Portugal
Goreti Marreiros	ISEP/IPP-GECAD, Portugal
Angel Martin Del Rey	Department of Applied Mathematics, Universidad de Salamanca, Spain
Ester Martinez-Martin	Universidad de Alicante, Spain
Philippe Mathieu	University of Lille 1, France
Kenji Matsui	Osaka Institute of Technology, Japan
Shimpei Matsumoto	Hiroshima Institute of Technology, Japan
Rene Meier	Lucerne University of Applied Sciences, Switzerland
José Ramón Méndez Reboredo	University of Vigo, Spain
Mohd Saberi Mohamad	United Arab Emirates University, United Arab Emirates
Jose M. Molina	Universidad Carlos III de Madrid, Spain
Miguel Molina-Solana	Data Science Institute - Imperial College London, UK
Stefania Monica	Universita' degli Studi di Parma, Italy
Naoki Mori	Osaka Prefecture University, Japan
Paulo Moura Oliveira	UTAD University, Portugal
Paulo Mourao	University of Minho, Portugal
Muhammad Marwan Muhammad Fuad	Technical University of Denmark, Denmark
Antonio J.R. Neves	University of Aveiro, Portugal
Jose Neves	University of Minho, Portugal
Julio Cesar Nievola	Pontifícia Universidade Católica do Paraná - PUCPR Programa de Pós Graduação em Informática Aplicada, Brazil
Nadia Nouali-Taboudjemat	CERIST, Algeria
Paulo Novais	University of Minho, Portugal
José Luis Oliveira	University of Aveiro, Portugal
Tiago Oliveira	National Institute of Informatics, Japan

Sigeru Omatu	Hiroshima University, Japan
Mauricio Orozco-Alzate	Universidad Nacional de Colombia, Colombia
Sascha Ossowski	University Rey Juan Carlos, Spain
Miguel Angel Patricio	Universidad Carlos III de Madrid, Spain
Juan Pavón	Universidad Complutense de Madrid, Spain
Reyes Pavón	University of Vigo, Spain
Pawel Pawlewski	Poznan University of Technology, Poland
Stefan-Gheorghe Pentiuc	University Stefan cel Mare Suceava, Romania
Antonio Pereira	Escola Superior de Tecnologia e Gestão do IPLeiria, Portugal
Tiago Pinto	Polytechnic of Porto, Portugal
Julio Ponce	Universidad Autónoma de Aguascalientes, Mexico
Juan-Luis Posadas-Yague	Universitat Politècnica de València, Spain
Jose-Luis Poza-Luján	Universitat Politècnica de València, Spain
Isabel Praça	GECAD/ISEP, Portugal
Radu-Emil Precup	Politehnica University of Timisoara, Romania
Mar Pujol	Universidad de Alicante, Spain
Francisco A. Pujol	Specialized Processor Architectures Lab, DTIC, EPS, University of Alicante, Spain
Araceli Queiruga-Dios	Department of Applied Mathematics, Universidad de Salamanca, Spain
Mariano Raboso Mateos	F Consejería de Educación - Junta de Andalucía, Spain
Miguel Rebollo	Universitat Politècnica de València, Spain
Manuel Resinas	University of Seville, Spain
Jaime A. Rincon	Universitat Politècnica de València, Spain
Ramon Rizo	Universidad de Alicante, Spain
Sergi Robles	Universitat Autònoma de Barcelona, Spain
Sara Rodríguez	University of Salamanca, Spain
Iván Rodríguez Conde	University of Arkansas at Little Rock, EE.UU
Cristian Aaron Rodriguez Enriquez	Instituto Tecnológico de Orizaba, México
Luiz Romao	Univille, Mexico
Gustavo Santos-Garcia	Universidad de Salamanca, Spain
Ichiro Satoh	National Institute of Informatics, Japan
Yann Secq	Université Lille I, France
Ali Selamat	Universiti Teknologi Malaysia, Malaysia
Emilio Serrano	Universidad Politécnica de Madrid, Spain
Mina Sheikhalishahi	Consiglio Nazionale delle Ricerche, Italy
Amin Shokri Gazafroudi	Universidad de Salamanca, Spain
Fábio Silva	University of Minho, Portugal
Nuno Silva	DEI & GECAD - ISEP - IPP, Portugal
Paweł Sitek	Kielce University of Technology, Poland
Pedro Sousa	University of Minho, Portugal

Richard Souza	UTFPR, Brazil
Shudong Sun	Northwestern Polytechnical University, China
Masaru Teranishi	Hiroshima Institute of Technology, Japan
Adrià Torrens Urrutia	Universitat Rovira i Virgili, Spain
Leandro Tortosa	University of Alicante, Spain
Volodymyr Turchenko	Research Institute for Intelligent Computing Systems, Ternopil National Economic University, Ucrania
Miki Ueno	Toyohashi University of Technology, Japan
Zita Vale	GECAD - ISEP/IPP, Portugal
Rafael Valencia-Garcia	Departamento de Informática y Sistemas. Universidad de Murcia, Spain
Miguel A. Vega-Rodríguez	University of Extremadura, Spain
Maria João Viamonte	Instituto Superior de Engenharia do Porto, Portugal
Paulo Vieira	Insituto Politécnico da Guarda, Portugal
José Ramón Villar	University of Oviedo, Spain
Friederike Wall	Alpen-Adria-Universitaet Klagenfurt, Austria
Zhu Wang	XINGTANG Telecommunications Technology Co., Ltd., China
Li Weigang	University of Brasilia, Brazil
Bozena Wozna-Szczesniak	Institute of Mathematics and Computer Science, Jan Dlugosz University in Czestochowa
Michal Wozniak	Wroclaw University of Technology, Poland
Takuya Yoshihiro	Faculty of Systems Engineering, Wakayama University, Japan
Michifumi Yoshioka	Osaka Pref. Univ., Japan
Andrzej Zbrzezny	Institute of Mathematics and Computer Science, Jan Dlugosz University in Czestochowa, Poland
Agnieszka Zbrzezny	Institute of Mathematics and Computer Science, Jan Dlugosz, Poland University in Czestochowa, Poland
Omar Zermeno	Monterrey Tech, Mexico
Zhen Zhang	Dalian University of Technology, China
Hengjie Zhang	Hohai University, China
Shenghua Zhou	Xidian University, China
Yun Zhu	Shaanxi Normal University, China
Andre Zúquete	University of Aveiro, Portugal

Organizing Committee

Juan M. Corchado Rodríguez	University of Salamanca, Spain/AIR Institute, Spain
Fernando De la Prieta	University of Salamanca, Spain

Sara Rodríguez González	University of Salamanca, Spain
Javier Prieto Tejedor	University of Salamanca, Spain/AIR Institute, Spain
Pablo Chamoso Santos	University of Salamanca, Spain
Belén Pérez Lancho	University of Salamanca, Spain
Ana Belén Gil González	University of Salamanca, Spain
Ana De Luis Reboredo	University of Salamanca, Spain
Angélica González Arrieta	University of Salamanca, Spain
Emilio S. Corchado Rodríguez	University of Salamanca, Spain
Angel Luis Sánchez Lázaro	University of Salamanca, Spain
Alfonso González Briones	University of Salamanca, Spain
Yeray Mezquita Martín	University of Salamanca, Spain
Javier J. Martín Limorti	University of Salamanca, Spain
Alberto Rivas Camacho	University of Salamanca, Spain
Ines Sitton Candanedo	University of Salamanca, Spain
Elena Hernández Nieves	University of Salamanca, Spain
Beatriz Bellido	University of Salamanca, Spain
María Alonso	University of Salamanca, Spain
Diego Valdeolmillos	AIR Institute, Spain
Roberto Casado Vara	University of Salamanca, Spain
Sergio Marquez	University of Salamanca, Spain
Jorge Herrera	University of Salamanca, Spain
Marta Plaza Hernández	University of Salamanca, Spain
Guillermo Hernández González	University of Salamanca, Spain/AIR Institute, Spain
Ricardo S. Alonso Rincón	AIR Institute, Spain
Javier Parra	University of Salamanca, Spain

DCAI 2021 Sponsors

Sponsors

IBM indra

Organizers

OSAKA INSTITUTE OF TECHNOLOGY

UNIVERSIDAD DE GRANADA

HIROSHIMA UNIVERSITY

QUT Queensland University of Technology

VNIVERSIDAD DSALAMANCA

Support from National Associations

© Copyright 2021

Contents

A Theorem Proving Approach to Formal Verification of a Cognitive Agent

Alexander Birch Jensen(✉)

DTU Compute - Department of Applied Mathematics and Computer Science, Technical University of Denmark, Richard Petersens Plads, Building 324, 2800 Kongens Lyngby, Denmark
aleje@dtu.dk

Abstract. Theorem proving approaches have successfully been applied to verify various traditional software and hardware systems. However, it has not been explored how theorem proving can be applied to verify agent systems. We formalize a framework for verification of cognitive agent programs in a proof assistant. This enables access to powerful automation and provides assurance that our results are correct.

1 Introduction

Demonstrating reliability plays a central role in the development and deployment of software systems in general. For cognitive multi-agent systems (CMAS) we observe particularly complex behaviour patterns often exceeding those of procedural programs [21]. This calls for techniques that are specially tailored towards demonstrating the reliability of cognitive agents. CMAS are systems consisting of agents that incorporate cognitive concepts such as beliefs and goals. The engineering of these systems is facilitated by dedicated programming languages that operate with high-level cognitive concepts, thus enabling compact representation of complex decision-making mechanisms.

The present paper applies theorem proving to formalize a verification framework for the agent programming language GOAL [8,9] in a proof assistant—a software tool that assists the user in the development of formal proofs. State-of-the-art proof assistants have proven successful in verifying various software and hardware systems [18]. The formalization is based on the work of [3] and developed in the higher-order logic proof assistant Isabelle/HOL [17]. The expected outcome is twofold: firstly, the automation of the proof assistant can be exploited to assist in the verification process; secondly, we gain assurance that any agent proof is correct as they are based on the formal semantics of GOAL. We identify as our first major milestone: verify a GOAL agent that solves an instance of a Blocks World for Teams problem [14].

The present paper is a substantially extended and revised version of our short paper, in the student session with no proceedings, at EMAS 2021 (9th International Workshop on Engineering Multi-Agent Systems): Formal Verification of a Cognitive Agent Using Theorem Proving.

© The Author(s), under exclusive license to Springer Nature Switzerland AG 2022
K. Matsui et al. (Eds.): DCAI 2021, LNNS 327, pp. 1–11, 2022.
https://doi.org/10.1007/978-3-030-86261-9_1

The Isabelle files are publicly available online:

https://people.compute.dtu.dk/aleje/public/

The Isabelle/HOL formalization is around 2500 lines of code in total and loads in less than 10 s on a modern laptop.

The paper is structured as follows. Section 2 considers related work. Sections 3, 4, 5, 6 and 7 describe the details of our formalization. Finally, Sect. 8 concludes.

2 Related Work

This paper expands on ideas from our work on verification of GOAL agents. In [13], we first sketched how to transform GOAL agent code into an agent logic that enabled its verification. We further expanded on these ideas in [10], and in [11,12] we argued for the use of theorem proving to verify CMAS.

We have seen practical tools that demonstrate reliability of CMAS using a model checking approach such as by [4,15]. The former suggests to integrate a model checker on top of the program interpreter. Model checking draws many parallels to our approach. For instance, the properties to be checked are usually formulated in temporal logic. However, a noticeable difference is how the property is verified. Using theorem proving, we are not explicitly checking all states of the system. Another dominant approach to verification of agent systems is through various testing methods, such as by [6,16]. The former proposes an automated testing framework that automatically detects failures in a cognitive agent system. For formal verification we have seen some work that applies theorem proving such as by [1]. In particular, [19] explores verification of agent specifications. However, this work is mostly on the specification level and does not connect well with agent programming. Finally, [20] proposes to combine testing and formal verification as neither practically succeeds on its own in a complete demonstration of reliability.

In [5], a recent survey of logic-based technologies that also accounts for means of verification, we observe a high representation of model checking. Meanwhile, we do not find any mention of theorem proving nor proof assistants. This indicates that the MAS community has not adopted these methodologies and tools. The effectiveness of model checking techniques manifested itself at the time MAS gained traction which presumably contributed towards its popularity. While initial work on BDI agent logics showed good promise, their practical applications were not further explored. At this time, the automatic tools to reduce the verification effort were not as well established. Furthermore, the logic was very complex. However, [7] has since shown that such a complex logic is not required.

3 Logic Framework

Before we get started on the formalization of the GOAL agent programming language, and its verification framework, we will set up a general framework for classical logic without quantifiers. We call it a framework because it can be instantiated for any type of atom:

datatype $'a\ \Phi_P = Atom\ 'a \mid Negation\ \langle 'a\ \Phi_P \rangle\ (\langle\neg\rangle) \mid$
$\quad Implication\ \langle 'a\ \Phi_P\rangle\ \langle 'a\ \Phi_P\rangle$ (**infixr** $\langle\longrightarrow\rangle$ 60) $\mid$
$\quad Disjunction\ \langle 'a\ \Phi_P\rangle\ \langle 'a\ \Phi_P\rangle$ (**infixl** $\langle\vee\rangle$ 70) $\mid$
$\quad Conjunction\ \langle 'a\ \Phi_P\rangle\ \langle 'a\ \Phi_P\rangle$ (**infixl** $\langle\wedge\rangle$ 80)

The type variable $'a$ signifies an arbitrary type. For classical propositional logic, as we have come to know it, we would instantiate with the type for strings or natural numbers.

The semantics of our logic is given by an interpretation (a mapping from atoms to truth values):

primrec $semantics_P :: \langle('a \Rightarrow bool) \Rightarrow\ 'a\ \Phi_P \Rightarrow bool\rangle$ **where**
$\langle semantics_P\ f\ (Atom\ x) = f\ x\rangle \mid$
$\langle semantics_P\ f\ (\neg\ p) = (\neg semantics_P\ f\ p)\rangle \mid$
$\langle semantics_P\ f\ (p \longrightarrow q) = (semantics_P\ f\ p \longrightarrow semantics_P\ f\ q)\rangle \mid$
$\langle semantics_P\ f\ (p \vee q) = (semantics_P\ f\ p \vee semantics_P\ f\ q)\rangle \mid$
$\langle semantics_P\ f\ (p \wedge q) = (semantics_P\ f\ p \wedge semantics_P\ f\ q)\rangle$

These ideas are very much adapted from the work of [2] that uses a similar technique in the definitions of syntax and semantics, although for a syntax which includes quantifiers.

We define the entailment relation for sets of formulas of both sides:

abbreviation $entails :: \langle 'a\ \Phi_P\ set \Rightarrow\ 'a\ \Phi_P\ set \Rightarrow bool\rangle$ (**infix** $\langle\models_P\#\rangle$ 50) **where**
$\langle\Gamma \models_P\# \Delta \equiv (\forall f.\ (\forall p{\in}\Gamma.\ semantics_P\ f\ p) \longrightarrow (\exists p{\in}\Delta.\ semantics_P\ f\ p))\rangle$

For derivation of formulas, we define a standard sequent calculus inductively:

inductive $seqc :: \langle 'a\ \Phi_P\ multiset \Rightarrow\ 'a\ \Phi_P\ multiset \Rightarrow bool\rangle$ (**infix** $\langle\vdash_P\rangle$ 40) **where**
$Axiom$: $\langle\{\ p\ \} + \Gamma \vdash_P \Delta + \{\ p\ \}\rangle \mid$
$L\text{-}Neg$: $\langle\Gamma \vdash_P \Delta + \{\ p\ \} \Longrightarrow \Gamma + \{\ \neg\ p\ \} \vdash_P \Delta\rangle \mid$
$R\text{-}Neg$: $\langle\Gamma + \{\ p\ \} \vdash_P \Delta \Longrightarrow \Gamma \vdash_P \Delta + \{\ \neg\ p\ \}\rangle \mid$
$R\text{-}Imp$: $\langle\Gamma + \{\ p\ \} \vdash_P \Delta + \{\ q\ \} \Longrightarrow \Gamma \vdash_P \Delta + \{\ p \longrightarrow q\ \}\rangle \mid$
$R\text{-}Or$: $\langle\Gamma \vdash_P \Delta + \{\ p,\ q\ \} \Longrightarrow \Gamma \vdash_P \Delta + \{\ p \vee q\ \}\rangle \mid$
$L\text{-}And$: $\langle\Gamma + \{\ p,\ q\ \} \vdash_P \Delta \Longrightarrow \Gamma + \{\ p \wedge q\ \} \vdash_P \Delta\rangle \mid$
$R\text{-}And$: $\langle\Gamma \vdash_P \Delta + \{\ p\ \} \Longrightarrow \Gamma \vdash_P \Delta + \{\ q\ \} \Longrightarrow \Gamma \vdash_P \Delta + \{\ p \wedge q\ \}\rangle \mid$
$L\text{-}Or$: $\langle\Gamma + \{\ p\ \} \vdash_P \Delta \Longrightarrow \Gamma + \{\ q\ \} \vdash_P \Delta \Longrightarrow \Gamma + \{\ p \vee q\ \} \vdash_P \Delta\rangle \mid$
$L\text{-}Imp$: $\langle\Gamma \vdash_P \Delta + \{\ p\ \} \Longrightarrow \Gamma + \{\ q\ \} \vdash_P \Delta \Longrightarrow \Gamma + \{\ p \longrightarrow q\ \} \vdash_P \Delta\rangle$

Note that in the above we have suppressed the special syntax character #. It is used by Isabelle to distinguish between instances of regular sets and multisets.

We can state and prove a soundness theorem for the sequent calculus:

theorem $soundness_P$: $\langle\Gamma \vdash_P\# \Delta \Longrightarrow set\text{-}mset\ \Gamma \models_P\# set\text{-}mset\ \Delta\rangle$
by (*induct rule*: *sequent-calculus.induct*) *auto*

The proof is completed automatically by induction over the proof system.

Because the framework is generic, allowing for any type of atom, we can reuse much of the footwork as we consider particular instances. In the sequel, we will generally recognize the different instances by their subscript. For instance, we have $\models_L$ and $\vdash_L$ for propositional logic with natural numbers as atoms.

4 Cognitive Agents

This section marks the start of our formalization of GOAL. The cognitive capabilities of GOAL agents are facilitated by their cognitive states (beliefs and goals). A mental state consists of a belief and goal base, respectively:

type-synonym $mst = \langle(\Phi_L\ set \times \Phi_L\ set)\rangle$

Not all elements of the simple type *mst* qualify as actual mental states: a number of restrictions apply. We capture these by the following definition:

definition $is\text{-}mst :: \langle mst \Rightarrow bool\rangle\ (\langle\nabla\rangle)$ **where**
$\langle\nabla\ M \equiv let\ (\Sigma, \Gamma) = M\ in\ \neg\ \Sigma \models_L \bot_L \wedge (\forall\gamma\in\Gamma.\ \neg\ \Sigma \models_L \gamma \wedge \neg\ \{\} \models_L \neg\ \gamma)\rangle$

The definition states that the belief base (Σ) is consistent, no goals ($\gamma \in \Gamma$) of the agent are entailed by its beliefs, and that all goals are satisfiable.

The belief and goal operators enable the agent's introspective properties:

fun $semantics_M{}' :: \langle mst \Rightarrow Atoms_M \Rightarrow bool\rangle$ **where**
$\langle semantics_M{}'\ (\Sigma, \text{-})\ (Bl\ \Phi) = (\Sigma \models_L \Phi)\rangle\ |$
$\langle semantics_M{}'\ (\Sigma, \Gamma)\ (Gl\ \Phi) = (\neg\ \Sigma \models_L \Phi \wedge (\exists\gamma\in\Gamma.\ \{\} \models_L \gamma \longrightarrow \Phi))\rangle$

The type $Atoms_M$ is for the atomic formulas. The belief operator succeeds if the queried formula is entailed by the belief base. The goal operator succeeds if a formula in the goal base entails it (i.e. is a subgoal; note that a formula always entails itself) and if it is not entailed by the belief base. Mental state formulas emerge from Boolean combinations of these operators.

Alongside the semantics, we define a proof system for mental state formulas:

inductive $derive_M :: \langle\Phi_M \Rightarrow bool\rangle\ (\langle\vdash_M \text{-}\rangle\ 40)$ **where**
$R1$: $\langle\vdash_P \varphi \Longrightarrow \vdash_M \varphi\rangle\ |$
$R2$: $\langle\vdash_P \Phi \Longrightarrow \vdash_M (B\ \Phi)\rangle\ |$
$A1$: $\langle\vdash_M ((B\ \Phi \longrightarrow \psi) \longrightarrow (B\ \Phi) \longrightarrow (B\ \psi))\rangle\ |$
$A2$: $\langle\vdash_M (\neg\ (B\ \bot_L))\rangle\ |$
$A3$: $\langle\vdash_M (\neg\ (G\ \bot_L))\rangle\ |$
$A4$: $\langle\vdash_M ((B\ \Phi) \longrightarrow (\neg\ (G\ \Phi)))\rangle\ |$
$A5$: $\langle\vdash_P (\Phi \longrightarrow \psi) \Longrightarrow \vdash_M ((\neg\ (B\ \psi)) \longrightarrow (G\ \Phi) \longrightarrow (G\ \psi))\rangle$

The rule $R1$ states that any classical tautology is derivable. The rule $R2$ states that an agent believes any tautology. Lastly, $A1$–$A5$ state properties of the goal and belief operators, e.g. that B distributes over implication (A1).

We state and prove the soundness theorem for $\vdash_M$:

theorem $soundness_M$: **assumes** $\langle\nabla\ M\rangle$ **shows** $\langle\vdash_M \Phi \Longrightarrow M \models_M \Phi\rangle$

Many of the rules are sound due to the properties of mental states that can be inferred from the semantics and mental state definition. The proof obligations are too convoluted for Isabelle to automatically discharge them. The proof is rather extensive and has been omitted in the present paper. The proof is started by applying induction over the rules of $\vdash_M$, meaning that we prove the soundness of each rule.

5 Agent Capabilities

In this section, we introduce capabilities (actions) for agents alongside an agent definition. To this end, we enrich our logic to facilitate reasoning about enabledness of actions. Consequently, we need to extend both the proof system and semantics.

We start with a datatype for the different kinds of agent capabilities:

datatype *cap* = *basic Bcap* | *adopt* (*cget*: Φ_L) | *drop* (*cget*: Φ_L)

The first option takes an identifier *BCap* of a user-specified action (we have chosen to identify actions by natural numbers). The action *adopt* adds a formula to the goal base and *drop* removes all goals that entail the given formula.

We extend the notion of a basic action with that of a conditional action:

type-synonym *cond-act* = ‹Φ_M × *cap*›

Here, a condition (on the mental state of the agent) states when the action may be selected for execution; notation: $\varphi \triangleright$ *do a* for condition φ and basic action a.

Due to execution actions, the belief update capabilities of agents are defined by a function $\mathcal{T}$. Given an action identifier and a mental state, the result is an updated belief base. The update to the goal base, outside of the execution of the built-in GOAL actions *adopt* and *drop*, is inferred from the default commitment strategy in which goals are only dropped once believed to be true.

We instantiate a context in which, for a single agent, we assume the existence of a fixed $\mathcal{T}$, a set of conditional actions Π and an initial mental state M_0:

locale *single-agent* =
  fixes
    $\mathcal{T}$ :: *bel-upd-t* **and** Π :: ‹*cond-act set*› **and** M_0 :: *mst*
  assumes
    is-agent: ‹$\Pi \neq \{\} \wedge \nabla\ M_0$› **and**
    $\mathcal{T}$*-consistent*: ‹$(\exists \varphi.\ (\varphi,\ basic\ a) \in \Pi) \longrightarrow \neg\ \Sigma \models_L \bot_L \longrightarrow$
      $\mathcal{T}\ a\ (\Sigma, \Gamma) \neq None \longrightarrow \neg\ the\ (\mathcal{T}\ a\ (\Sigma, \Gamma)) \models_L \bot_L$› **and**
    $\mathcal{T}$*-in-domain*: ‹$\mathcal{T}\ a\ (\Sigma, \Gamma) \neq None \longrightarrow (\exists \varphi.\ (\varphi,\ basic\ a) \in \Pi)$›

Everything defined within a context will be local to its scope and will have those fixed variables available in definitions, proofs etc. An instance may gain access to the context by proving the assumptions true for a given set of input variables.

While the belief update capabilities are fixed, the effects on the goal base are defined by a function $\mathcal{M}$ which returns the resulting mental state after executing an action (as such it builds upon the function $\mathcal{T}$):

fun *mst-transformer* :: ‹*cap* ⇒ *mst* ⇒ *mst option*› (‹$\mathcal{M}$›) **where**
  ‹$\mathcal{M}\ (basic\ n)\ (\Sigma, \Gamma) = (case\ \mathcal{T}\ n\ (\Sigma, \Gamma)\ of$
    $Some\ \Sigma' \Rightarrow Some\ (\Sigma', \Gamma - \{\psi \in \Gamma.\ \Sigma' \models_L \psi\})$ | - ⇒ $None)$› |
  ‹$\mathcal{M}\ (drop\ \Phi)\ (\Sigma, \Gamma) = Some\ (\Sigma, \Gamma - \{\psi \in \Gamma.\ \{\psi\} \models_L \Phi\})$› |
  ‹$\mathcal{M}\ (adopt\ \Phi)\ (\Sigma, \Gamma) =$
    $(if\ \neg\ \{\} \models_L \neg\ \Phi \wedge \neg\ \Sigma \models_L \Phi\ then\ Some\ (\Sigma, \Gamma \cup \{\Phi\})\ else\ None)$›

The first case captures the default commitment strategy. The case for *drop* ϕ removes all goals that entail ϕ. Finally, the case for *adopt* ϕ adds the goal ϕ. The execution of actions gives rise to a notion of transitions between states:

definition *transition* :: ‹*mst* ⇒ *cond-act* ⇒ *mst* ⇒ *bool*› (‹- →- -›) **where**
‹$M \rightarrow b\ M' \equiv$ *let* $(\varphi, a) = b$ *in* $b \in \Pi \wedge M \models_M \varphi \wedge \mathcal{M}\ a\ M =$ *Some* M'›

If b is a conditional action, and the condition φ holds in M, then there is a possible transition between M and M' where M' is the result of $\mathcal{M}\ a\ M$.

In sequence, transitions form traces. A trace is an infinite interleaving of mental states and conditional actions:

codatatype *trace* = *Trace mst* ‹*cond-act* × *trace*›

Just as for mental states, we need a definition to capture the meaning of a trace:

definition *is-trace* :: ‹*trace* ⇒ *bool*› **where**
‹*is-trace* $s \equiv \forall i.$ (*let* $(M, M', (\varphi, a))$ = (*st-nth* s i, *st-nth* s $(i+1)$, *act-nth* s i) *in*
$(\varphi, a) \in \Pi \wedge ((M \rightarrow (\varphi, a)\ M') \vee \mathcal{M}\ a\ M =$ *None* $\wedge M = M'))$›

For all i there is a transition between M_i (the i'th state of the trace) and M_{i+1} due to an action $\varphi \triangleright$ *do* a, or the action is not enabled and $M_{i+1} = M$.

As such, a trace describes a possible execution sequence. In a fair trace each of the actions is scheduled infinitely often:

definition *fair-trace* :: ‹*trace* ⇒ *bool*› **where**
‹*fair-trace* $s \equiv \forall\ b \in \Pi\ .\ \forall i\ .\ \exists\ j > i.$ *act-nth* $s\ j = b$›

We define an agent as the set of fair traces starting from the initial mental state:

definition *Agent* :: ‹*trace set*› **where**
‹*Agent* $\equiv \{s$. *is-trace* $s \wedge$ *fair-trace* $s \wedge$ *st-nth* s $0 = M_0\}$›

We now return to our mental state logic and define the semantics of enabledness:

‹*semantics*$_E$' M (*enabled-basic* a) = $(\mathcal{M}\ a\ M \neq$ *None*)› |
‹*semantics*$_E$' M (*enabled-cond* b) = $(\exists M'. (M \rightarrow b\ M'))$›

A basic action is enabled if the mental state transformer $\mathcal{M}$ is defined. For a conditional action, we may express it as the existence of a transition from M.

We also extend the proof system with additional proof rules:

inductive *provable*$_E$ :: ‹Φ_E ⇒ *bool*› (‹$\vdash_E$ -› *40*) **where**
R1: ‹$\vdash_P \varphi \Longrightarrow \vdash_E \varphi$› |
R$_M$: ‹$\vdash_M \varphi \Longrightarrow \vdash_E (\varphi^E)$› |
E1: ‹$\vdash_P \varphi \Longrightarrow \vdash_E$ (*enabledb* a) $\Longrightarrow (\varphi \triangleright$ *do* $a) \in \Pi \Longrightarrow \vdash_E$ (*enabled* $(\varphi \triangleright$ *do* a))› |
E2: ‹$\vdash_E$ (*enabledb* (*drop* Φ))› |
R3: ‹$\neg \vdash_P \neg\ \Phi \Longrightarrow \vdash_E (\neg\ ((B\ \Phi)^E) \longleftrightarrow$ (*enabledb* (*adopt* Φ)))› |
R4: ‹$\vdash_P (\neg\ \Phi) \Longrightarrow \vdash_E (\neg$ (*enabledb* (*adopt* Φ)))› |
R5: ‹$\forall M.\ \mathcal{T}\ a\ M \neq$ *None* $\Longrightarrow \vdash_E$ (*enabledb* (*basic* a))›

The use of φ^E is merely for converting a formula φ to a more expressive datatype—the structure of the formulas is preserved. It can be disregarded for the purpose of this paper. The new proof rules mainly state properties of formulas concerning enabledness of actions while $R1$ and R_M transfer rules from $\vdash_P$ and $\vdash_M$, respectively.

We state the soundness theorem for our extended proof system:

theorem *soundness*$_E$: **assumes** ‹$\nabla\ M$› **shows** ‹$\vdash_E \varphi \Longrightarrow M \models_E \varphi$›

The proof has been omitted in the present paper.

6 Hoare Logic for Actions

To facilitate reasoning about the effects (and non-effects) of actions, we introduce a specialized form of Hoare logic in which Hoare triples state pre- and postconditions for actions. The following datatype is for basic and conditional action Hoare triples, respectively:

datatype *hoare-triple* = *htb* (*pre*: Φ_M) *cap* (*post*: Φ_M) | *htc* Φ_M *cond-act* Φ_M

Let us introduce some notation: $\varphi\,[s\ i]$ means that the formula φ is evaluated in the i'th state of trace s, and $\{\ \varphi\ \}\ a\ \{\ \psi\ \}$ is a Hoare triple for action a with precondition φ and postcondition ψ.

We now define the semantics of Hoare triples:

fun *semantics*$_H$ :: ‹*hoare-triple* ⇒ *bool*› (‹$\models_H$›) **where**
‹$\models_H$ **{** φ **}** a **{** ψ **}** = $(\forall M.\ \nabla M \longrightarrow$
$(M \models_E (\varphi^E) \wedge\ (enabledb\ a) \longrightarrow the\ (\mathcal{M}\ a\ M) \models_M \psi) \wedge$
$(M \models_E (\varphi^E) \wedge \neg(enabledb\ a) \longrightarrow M \models_M \psi))$› |
‹$\models_H$ **{** φ **}** $(\upsilon \triangleright do\ b)$ **{** ψ **}** =
$(\forall\ s \in Agent.\ \forall i.\ ((\varphi[s\ i]_M) \wedge (\upsilon \triangleright do\ b) = (act\text{-}nth\ s\ i) \longrightarrow (\psi[s\ (i{+}1)]_M)))$›

The first case, for basic actions, states: for all mental states, if the precondition holds in M and the action is enabled then the postcondition should hold in the successor state. Otherwise, if the precondition holds and the action is not enabled then the precondition should hold in the current state. For conditional actions, the definition takes a different form, but essentially captures the same meaning except that the condition υ must also hold in M.

We round out this section with a lemma showing the relation between Hoare triples for basic actions and Hoare triples for conditional actions:

lemma *hoare-triple-cond-from-basic*:
assumes ‹$\models_H$ **{** $\varphi \wedge \psi$ **}** a **{** φ' **}**›
and ‹$\forall\ s \in Agent.\ \forall i.\ st\text{-}nth\ s\ i \models_M (\varphi \wedge \neg\psi) \longrightarrow \varphi'$›
shows ‹$\models_H$ **{** φ **}** $(\psi \triangleright do\ a)$ **{** φ' **}**›

The proof has been omitted in the present paper.

7 Specifying Agent Programs

We are now concerned with finding a proof system for Hoare triples. Clearly, such a proof system depends on the agent program specification. We therefore first describe how to specify agent programs. We define an agent specification to consist of a set of Hoare triples (that the user claims to be true) and predicates for enabledness of actions. In the following, the type *ht-spec* represents an agent specification.

In the prequel, we relied on a fixed function for the agent's belief update capabilities. This does not work well in practice. We would rather specify agent programs by means of Hoare triples. In order to link a specification to a function for belief update capabilities, we first need to ensure that the specification is satisfiable. In other words, that the specified Hoare triples are not contradictory.

We are not immediately interested in actually computing the semantics of Hoare triples for a given $\mathcal{T}$. Luckily, it suffices for us to prove the existence of a $\mathcal{T}$ (model) that complies with our specification. Proving this existence allows for the introduction of a $\mathcal{T}$ due to Hilbert's Axiom of Choice.

We need to define compliance for a single Hoare triple:

fun *complies-ht* :: ‹*mst* ⇒ *bel-upd-t* ⇒ Φ_M ⇒ (Φ_M × *Bcap* × Φ_M) ⇒ *bool*› **where**
‹*complies-ht* $M\ \mathcal{T}\ \Phi\ (\varphi,\ n,\ \psi) =$
$((M \models_M \Phi \longleftrightarrow \mathcal{T}\ n\ M \neq \mathit{None}) \wedge$
$(\neg\ (\mathit{fst}\ M) \models_L \bot_L \longrightarrow \mathcal{T}\ n\ M \neq \mathit{None} \longrightarrow \neg \mathit{the}\ (\mathcal{T}\ n\ M) \models_L \bot_L) \wedge$
$(M \models_M \varphi \wedge M \models_M \Phi \longrightarrow \mathit{the}\ (\mathcal{M}\ (\mathit{basic}\ n)\ M) \models_M \psi) \wedge$
$(M \models_M \varphi \wedge M \models_M \neg\ \Phi \longrightarrow M \models_M \psi))$›

The definition is inferred from the semantics of Hoare triples, the definition of enabledness and lastly from a consistency property on $\mathcal{T}$. The specification complies when all Hoare triples comply simultaneously.

The following lemma states that proving the existence of a model can be achieved by proving the model existence for each action separately:

lemma *model-exists-disjoint*:
assumes ‹*is-ht-spec S*› **and** ‹$\forall\ s{\in}\mathit{set}\ S.\ \exists \mathcal{T}.\ \mathit{complies'}\ s\ \mathcal{T}$›
shows ‹$\exists \mathcal{T}.\ \mathit{complies}\ S\ \mathcal{T}$›

The lemma above forms the basis for the proof of a model existence lemma:

lemma *model-exists*: ‹$\mathit{is\text{-}ht\text{-}spec}\ S \Longrightarrow \exists \mathcal{T}.\ \mathit{complies}\ S\ \mathcal{T}$›

Here, the definition *is-ht-spec* states that the agent specification S is valid; most notably, it is satisfiable. The expression $\exists \mathcal{T}.$ *complies* $S\ T$ states that there exists a model that S complies with. We skip this definition, but note that it is based on the definition of compliance for Hoare triples that we hinted at previously.

We now extend the context to also fix a valid specification S that complies with our $\mathcal{T}$. In this context, we define a proof system for Hoare triples:

inductive derive_H :: ‹*hoare-triple* ⇒ *bool*› (‹$\vdash_H$›) **where**
import: ‹$(n,\ \Phi,\ \mathit{hts}) \in \mathit{set}\ S \Longrightarrow \{\ \varphi\ \}\ (\mathit{basic}\ n)\ \{\ \psi\ \} \in \mathit{set}\ \mathit{hts} \Longrightarrow$

$\vdash_H \{ \varphi \} \; (basic\ n) \; \{ \psi \}\rangle \mid$
persist: $\langle \neg\ is\text{-}drop\ a \Longrightarrow \vdash_H \{ (G\ \Phi) \ \} \; a \; \{ (B\ \Phi) \lor (G\ \Phi) \}\rangle \mid$
inf: $\langle \vdash_E ((\varphi^E) \longrightarrow \neg(enabledb\ a)) \Longrightarrow \vdash_H \{ \varphi \} \; a \; \{ \varphi \}\rangle \mid$
dropNegG: $\langle \vdash_H \{ \neg(G\ \Phi) \} \; (drop\ \psi) \; \{ \neg(G\ \Phi) \}\rangle \mid$
dropGCon: $\langle \vdash_H \{ \neg(G\ (\Phi \land \psi)) \land (G\ \Phi) \} \; (drop\ \psi) \; \{ G\ \Phi \}\rangle \mid$
rCondAct: $\langle \vdash_H \{ \varphi \land \psi \} \; a \; \{ \varphi' \} \Longrightarrow \vdash_M (\varphi \land \neg\psi) \longrightarrow \varphi' \Longrightarrow$
$\vdash_H \{ \varphi \} \; (\psi \rhd do\ a) \; \{ \varphi' \}\rangle \mid$
rImp: $\langle \vdash_M \varphi' \longrightarrow \varphi \Longrightarrow \vdash_H \{ \varphi \} \; a \; \{ \psi \} \Longrightarrow \vdash_M \psi \longrightarrow \psi' \Longrightarrow$
$\vdash_H \{ \varphi' \} \; a \; \{ \psi' \}\rangle \mid$
rCon: $\langle \vdash_H \{ \varphi_1 \} \; a \; \{ \psi_1 \} \Longrightarrow \vdash_H \{ \varphi_2 \} \; a \; \{ \psi_2 \} \Longrightarrow$
$\vdash_H \{ \varphi_1 \land \varphi_2 \} \; a \; \{ \psi_1 \land \psi_2 \}\rangle \mid$
rDis: $\langle \vdash_H \{ \varphi_1 \} \; a \; \{ \psi \} \Longrightarrow \vdash_H \{ \varphi_2 \} \; a \; \{ \psi \} \Longrightarrow \vdash_H \{ \varphi_1 \lor \varphi_2 \} \; a \; \{ \psi \}\rangle$

Note that a few rules have been left out from the present paper due to space limitations.

Because of the satisfiability of the specification, we can prove $\vdash_H$ sound:

theorem $soundness_H$: $\langle \vdash_H H \Longrightarrow \models_H H\rangle$

The proof has been omitted in the present paper.

This marks the end of our formalization. Work is ongoing on a temporal logic which facilitates stating properties of agent programs. As described by [3], correctness properties can be proved using the system above.

8 Concluding Remarks

We have argued that the reliability of agent systems plays a central role during their development and deployment. We have further pointed out the opportunity for a theorem proving approach to their formal verification.

The present paper has presented a formalization of a verification framework for agents of the GOAL agent programming language. The formalization is presented as a step-wise construction of the formal semantics and corresponding proof systems. Our current theory development still lacks a temporal logic layer that enables reasoning across states of the program, and thus facilitates stating properties concerning execution of the program. For instance, that from the initial mental state the agent reaches some state in which it believes its goals to be achieved. Ongoing work shows good promise on this front, but it is too early to share any results yet.

Further down the road, we need to devote attention to the limitations of the framework itself. For instance, we only consider single agents and deterministic environments, and we use a logic without quantifiers. These limitations call for non-trivial improvements and extensions. We should also note that the formalization of GOAL is not complete in the sense that some pragmatic aspects are not included such as dividing code into modules and communication between multiple agents.

The current progress shows good promise for a theorem proving approach using the Isabelle/HOL proof assistant. We find that its higher-order logic capabilities for programming and proving are sufficient to formalize GOAL effectively, at least up to this point in development. In conclusion, the main contribution of this paper is towards building a solid foundation for verification of agent programs that can exploit the powerful automation of proof assistants and provide assurance that results are trustworthy.

References

1. Alechina, N., Dastani, M., Khan, A.F., Logan, B., Meyer, J.J.: Using theorem proving to verify properties of agent programs. In: Dastani, M., Hindriks, K., Meyer, J.J. (eds.) Specification and Verification of Multi-agent Systems. pp. 1–33, Springer, Boston (2010). https://doi.org/10.1007/978-1-4419-6984-2_1
2. Berghofer, S.: First-order logic according to fitting. Archive of Formal Proofs (2007). Formal proof development. https://isa-afp.org/entries/FOL-Fitting.html
3. de Boer, F.S., Hindriks, K.V., van der Hoek, W., Meyer, J.J.: A verification framework for agent programming with declarative goals. J. Appl. Log. **5**, 277–302 (2007)
4. Bordini, R., Fisher, M., Wooldridge, M., Visser, W.: Model checking rational agents. IEEE Intell. Syst. **19**, 46–52 (2004)
5. Calegari, R., Ciatto, G., Mascardi, V., Omicini, A.: Logic-based technologies for multi-agent systems: a systematic literature review. Auton. Agents Multi-agent Syst. **35** (2020)
6. Dastani, M., Brandsema, J., Dubel, A., Meyer, J.J.: Debugging BDI-based multi-agent programs. In: Braubach, L., Briot, J.P., Thangarajah, J. (eds.) ProMAS 2009. LNCS, vol. 5919, pp. 151–169. Springer, Heidelberg (2010). https://doi.org/10.1007/978-3-642-14843-9_10
7. Hindriks, K., van der Hoek, W.: GOAL agents instantiate intention logic. In: Artikis, A., Craven, R., Kesim, C.N., Sadighi, B., Stathis, K. (eds.) Logic Programs, Norms and Action. LNCS, vol. 7360, pp. 196–219. Springer, Heidelberg (2008). https://doi.org/10.1007/978-3-642-29414-3_11
8. Hindriks, K.V.: Programming rational agents in GOAL. In: El Fallah Seghrouchni, A., Dix, J., Dastani, M., Bordini, R. (eds.) Multi-agent Programming, pp. 119–157. Springer, Boston (2009). https://doi.org/10.1007/978-0-387-89299-3_4
9. Hindriks, K.V., Dix, J.: GOAL: a multi-agent programming language applied to an exploration game. In: Shehory, O., Sturm, A. (eds.) Agent-Oriented Software Engineering, pp. 235–258. Springer, Heidelberg (2014). https://doi.org/10.1007/978-3-642-54432-3_12
10. Jensen, A.: Towards verifying a blocks world for teams GOAL agent. In: Rocha, A., Steels, L., van den Herik, J. (eds.) ICAART 2021, vol. 1, pp. 337–344. Science and Technology Publishing, New York (2021)
11. Jensen, A.: Towards verifying GOAL agents in Isabelle/HOL. In: Rocha, A., Steels, L., van den Herik, J. (eds.) ICAART 2021, vol. 1, pp. 345–352. Science and Technology Publishing, New York (2021)
12. Jensen, A., Hindriks, K., Villadsen, J.: On using theorem proving for cognitive agent-oriented programming. In: Rocha, A., Steels, L., van den Herik, J. (eds.) ICAART 2021, vol. 1, pp. 446–453. Science and Technology Publishing, New York (2021)
13. Jensen, A.B.: A verification framework for GOAL agents. In: EMAS 2020 (2020)
14. Johnson, M., Jonker, C., Riemsdijk, B., Feltovich, P.J., Bradshaw, J.: Joint activity testbed: blocks world for teams (BW4T). In: Aldewereld, H., Dignum, V., Picard, G. (eds.) ESAW 2009. LNCS, vol. 5881, pp. 254–256. Springer, Heidelberg (2009). https://doi.org/10.1007/978-3-642-10203-5_26

15. Jongmans, S.S., Hindriks, K., Riemsdijk, M.: Model checking agent programs by using the program interpreter. In: Dix J., Leite, J., Governatori, G., Jamroga, W. (eds.) CLIMA 2010. LNCS, vol. 6245, pp. 219–237. Springer, Heidelberg (2010). https://doi.org/10.1007/978-3-642-14977-1_17
16. Koeman, V., Hindriks, K., Jonker, C.: Automating failure detection in cognitive agent programs. IJAOSE **6**, 275–308 (2018)
17. Nipkow, T., Paulson, L., Wenzel, M.: Isabelle/HOL—A Proof Assistant for Higher-Order Logic. LNCS, vol. 2283. Springer, Heidelberg (2002). https://doi.org/10.1007/3-540-45949-9
18. Ringer, T., Palmskog, K., Sergey, I., Gligoric, M., Tatlock, Z.: QED at large: a survey of engineering of formally verified software. Found. Trends® Program. Lang. **5**(2–3), 102–281 (2019)
19. Shapiro, S., Lespérance, Y., Levesque, H.J.: The cognitive agents specification language and verification environment for multiagent systems. In: AAMAS 2002, pp. 19–26. Association for Computing Machinery (2002)
20. Winikoff, M.: Assurance of agent systems: what role should formal verification play? In: Dastani, M., Hindriks, K.V., Meyer, J.J.C. (eds.) Specification and Verification of Multi-agent Systems, pp. 353–383. Springer, Boston (2010). https://doi.org/10.1007/978-1-4419-6984-2_12
21. Winikoff, M., Cranefield, S.: On the testability of BDI agent systems. JAIR **51**, 71–131 (2014)

Check for updates

Parallelization of the Poisson-Binomial Radius Distance for Comparing Histograms of n-grams

Ana-Lorena Uribe-Hurtado(✉) and Mauricio Orozco-Alzate

Departamento de Informática y Computación, Universidad Nacional de Colombia - Sede Manizales, km 7 vía al Magdalena, Manizales 170003, Colombia
{alhurtadou,morozcoa}@unal.edu.co

Abstract. Text documents are typically represented as *bag-of-words* in order to facilitate subsequent steps in their analysis and classification. Such a representation tends to be high-dimensional and sparse since, for each document, a histogram of its n-grams must be created by considering a global—and thereby large—vocabulary that is common to the whole collection of texts under consideration. A straightforward and powerful way to further process the documents is computing pairwise distances between their *bag-of-words* representations. A proper distance to compare histograms must be chosen, for instance the recently proposed Poisson-Binomial radius (PBR) distance which has shown to be very competitive in terms of accuracy but somehow computationally costly in contrast with other classic alternatives. We present a GPU-based parallelization of the PBR distance for alleviating the cost of comparing large histograms of n-grams. Our experiments were performed with publicly available datasets of n-grams and showed that speed-ups between 12 and 17 times can be achieved with respect to the sequential implementation.

Keywords: GPU · Histograms · n-grams · Pairwise comparisons · Parallel computation · PBR distance

1 Introduction

Computing pairwise comparisons of documents is essential for subsequent analyses in natural language processing (NLP) such as, for instance, grouping texts by similarity or assigning them to a number of pre-defined classes. The quality of the comparison highly depends on the selection of an appropriate distance measure, which must be chosen according to the way the documents are represented. One of the paradigmatic ways of representing plain text documents is the so-called *bag-of-words* [2] strategy, either for individual words (also known as *1-grams*) or combinations of n of them (n*-grams*). The bag-of-words representation simply consists in considering the document as a collection of n-grams, together with the counts of how many times each n-gram appears in it. Said differently, for a given value of n, the bag-of-words representation of a document is just a histogram of its n-grams.

© The Author(s), under exclusive license to Springer Nature Switzerland AG 2022
K. Matsui et al. (Eds.): DCAI 2021, LNNS 327, pp. 12–21, 2022.
https://doi.org/10.1007/978-3-030-86261-9_2

The histograms derived from the bag-of-words representation are inherently high-dimensional, since they have as many bins as the number of different n-grams that may appear in a set of documents; that is, they are as long as the size of the global dictionary of the texts to be compared. In addition, the histograms tend to be sparse: i.e. they have many counts equal to zero, corresponding to the n-grams that do not occur in a given document but that are included in the considered dictionary. This scale of the histogram-based representations implies a great computing load for the subsequent steps of a NLP system and has even risen environmental concerns in the scientific community, as recently discussed in [8] for the case of deep neural networks. In spite of these apparent disadvantages, the bag-of-words representation is still widely used to represent text documents and has also been extended for representing objects of different nature such as audio signals [5], images [4] and, even, seismic events [1].

There are a number of distance measures that are specially designed to compare histograms; among them, the following are widely-applied: the Jeffrey's divergence, the χ^2 distance, the histogram intersection distance and the cross-entropy measure. Some of them require the histograms to be normalized, such that they can be interpreted as an estimation of the underlying probability mass function (PMF) [7]. Recently, Swaminathan and collaborators [9] proposed a novel measure to compare histograms called the *Poisson-Binomial radius (PBR)* distance, which they showed to outperform most of the other above-mentioned distances—in classification accuracy terms—when used in several problems of image classification. Such a performance superiority of the PBR distance was also confirmed in [6] for a problem of classifying many classes of plant leaves, whose images were preprocessed to be represented as texture histograms.

In spite of this advantage of the PBR distance, it was experimentally shown in [6] that its computational cost quickly surpasses those of the other competing distances as the length of the histogram grows. It is desirable, therefore, to design a parallel version of the PBR distance in order to alleviate its cost by taking advantage of the currently available multi-core and many-core computer architectures. The latter ones typically refer to the usage of graphics processing units (GPUs), which are particularly appropriate for the massive parallelization of simple individual computations.

According to this motivation and because we did not find in the literature neither an attempt of parallelizing the computation of the PBR distance nor an application of it to compare n-grams, we propose in this paper a GPU-based parallelization of the PBR distance for comparing histograms of n-grams. In particular, we test our proposal for comparing histograms of 1-grams and 2-grams with global dictionaries ranging from 59,717 bins—the smallest problem—to 4,219,523 bins—the largest one. The remaining part of this paper is organized as follows: the sequential version and the proposed parallel implementation of the PBR distance are presented in Sect. 2. The experimental evaluation in terms of elapsed times and speed-ups is shown in Sect. 3. Finally, our concluding remarks are given in Sect. 4.

2 Method

Consider that vectors $\mathbf{x}$ and $\mathbf{y}$ are used to store two histograms of length N each. The PBR distance between them is defined as follows [9]:

$$d_{PBR}(\mathbf{x}, \mathbf{y}) = \frac{\sum_{i=1}^{N} e_i(1 - e_i)}{N - \sum_{i=1}^{N} e_i}, \tag{1}$$

where,

$$e_i = x_i \ln\left(\frac{2x_i}{x_i + y_i}\right) + y_i \ln\left(\frac{2y_i}{x_i + y_i}\right) \tag{2}$$

The PBR distance is a semimetric because it does not obey the triangle inequality. However, it is still a proper distance because the indiscernibility and symmetry conditions are fulfilled. Notice also that the histograms $\mathbf{x}$ and $\mathbf{y}$ must be provided as proper PMFs (i.e. they must be normalized), such that neither negative nor undefined results appear when computing Eqs. (1) and (2), respectively. The sequential computation of the PBR distance is straightforward, as shown below.

2.1 Sequential Computation of the PBR Distance

The sequential procedure to compute the PBR distance is shown in Algorithm 1. Notice that conditionals are used to prevent an indetermination when computing the logarithms in Eq. 2. Notice also that there are no nested loops involved in the distance computation and, in consequence, the PBR distance is a bin-to-bin measure. Measures of this type are good candidates for parallel implementations due to the per-index independence of its computations before the summations.

2.2 Parallel Computation of the PBR Distance for GPU

For the sake of clarity and reproducibility, the parallel implementation of the PBR distance is presented here as snippets of CUDA C [3] code. The implementation consists of two kernel functions, both of them using global memory in GPU and its x dimension to execute the individual PBR operations. The first kernel, called PBRGPU (see Listing 1), uses the well-known solution to add vectors in GPU. The template function of the vector addition in GPU can be used because the computation of *part*1 and *part*2 of the PBR distance can be implemented in a bin-to-bin approach.

Algorithm 1. Sequential implementation to compute the PBR distance

procedure DISTPBR($\mathbf{x}, \mathbf{y}$) ▷ $\mathbf{x}$ and $\mathbf{y}$ are vectors containing the histograms
 $N \leftarrow dim(\mathbf{x})$ ▷ Histograms have N bins (N-dimensional vectors)
 $num \leftarrow 0; \quad partDen \leftarrow 0$ ▷ Initialize accumulators for summations
 for $i \leftarrow 1, \ldots, N$ **do** ▷ Loop through the entries of the vectors
  $part1 \leftarrow 0; \quad part2 \leftarrow 0$ ▷ Initialize default values for cases
  if $x_i \neq 0$ **then** ▷ First part of Eq. (2)
   $part1 \leftarrow x_i \ln\left(\frac{2x_i}{x_i+y_i}\right)$
  end if
  if $y_i \neq 0$ **then** ▷ Second part of Eq. (2)
   $part2 \leftarrow y_i \ln\left(\frac{2y_i}{x_i+y_i}\right)$
  end if
  $e \leftarrow part1 + part2$
  $num \leftarrow num + e(1-e)$ ▷ Summation in the numerator of Eq (1)
  $partDen \leftarrow partDen + e$ ▷ Summation in the denominator of Eq (1)
 end for
 return $\frac{num}{N-partDen}$ ▷ The PBR distance between $\mathbf{x}$ and $\mathbf{y}$
end procedure

The dimension, according to the number of threads per block in the grid, is defined as $int \quad blockSize = deviceProp.maxThreadsPerBlock$. Notice that we use all possible threads per block. The numbers of blocks in the grid is estimated by the expression in Eq. (3).

$$unsigned \quad int \quad gridSize = \frac{ceil(float)F}{blockSize} \tag{3}$$

The idx variable, which contains the position the current thread in each block of the GPU, is defined as shown in Eq. (4) and is invoked from a $\_\_global\_\_$ kernel. The GPU threads compute, concurrently, $part1$ and $part2$ variables from Algorithm 1 for the corresponding x_{idx} and y_{idx} bins of the histograms. Similarly, the results of each numerator and denominator positions are stored in two different vectors: $partNum\_d$ and $partDen\_d$, respectively. The computational complexity of this implementation in GPU, for two histograms fitting in memory, is $\mathcal{O}(1)$; however, as many threads as the smallest power of two greater than the length of the histograms are needed; see Eq. (3).

$$unsigned \quad int \quad idx = threadIdx.x + blockIdx.x \times blockDim.x \tag{4}$$

Listing 1. PBRGPU kernel function.

```
__device__ void PBRGPU(float *x_d, float *y_d, float *partNum, float *
        partDen, unsigned int idx)
{
  float part1     = 0;
  float part2     = 0;
  float e         = 0;

  //Calculates the first part of the PBR equation with each position indexed by
          idx in the vectors x_d and y_d
  if (x_d[idx] !=0)
  {
        float tem1 = (2.0 * x_d[idx])/(x_d[idx]+y_d[idx]);
        part1 = x_d[idx] * log(tem1);
  }

  //Calculates the second part of the PBR equation with each position indexed
          by idx in the vectors x_d and y_d
  if (y_d[idx] !=0)
  {
        float tem2 = (2.0 * y_d[idx])/(y_d[idx]+x_d[idx]);
        part2 = y_d[idx] * log(tem2);
   }
   e = part1 + part2;
   partNum[idx] = e * (1.0 - e); //Stores the numerator result of each idx
          position
   partDen[idx] = e; //Stores the numerator result of each idx position
}
```

Afterwards, the implementation adds the entries of arrays $partNum\_d$ and $partDen\_d$ using the well-known reduction algorithm, as explained in [3]. The second kernel function in GPU executes reduceNeighboredLessGPU (see Listing 2), to add $partNum\_d$ and $partDen\_d$. This kernel function uses 512 threads per block and calculates the grid size as $gridSize = \frac{(size+block.x-1)}{block.x}$. The size of each vector in GPU is the next power of two closest to the size of the input vector. The reduction strategy leaves the results of the sum of the threads per block in the first position (see line 24 in Listing 2) of each block of the vector pointers by $idata1$ and $idata2$. Finally, the kernel function leaves the results of the computation made by the threads of each block in the output vectors $opartNum\_d$ and $opartDen\_d$; each position of those vectors is indexed by $blockIdx.x$. In the output vectors $opartNum\_d$ and $opartDen\_d$ there are as many values as blocks that were added. These results are returned to the host in the vectors $opartNum\_h$ and $opartDen\_h$; finally, the total sum is made in CPU. The computational complexity of this algorithm is $\mathcal{O}(\log_2(N))$ where N is the length of the vector whose entries are added, which depends on the gridSize.

Listing 2. Kernel function for adding arrays via Reduction.

```
__device__ void reduceNeighboredLessGPU(float *partNum_d, float *
    opartNum_d, float *partDen_d, float *opartDen_d, unsigned int size)
{
  // set thread ID
  unsigned int tid = threadIdx.x;
  unsigned int idx = blockIdx.x * blockDim.x + threadIdx.x;
  // convert global data pointer to the local pointer of this block
  float *idata1 = partNum_d + blockIdx.x*blockDim.x;
  float *idata2 = partDen_d + blockIdx.x*blockDim.x;
  // boundary check
  if(idx >= size) return;
  // in-place reduction in global memory
  for (int stride = 1; stride < blockDim.x; stride *= 2) {
  // convert tid into local array index
    int index = 2 * stride * tid;
    if (index < blockDim.x) {
      idata1[index] += idata1[index + stride];
      idata2[index] += idata2[index + stride];
    }
    // synchronize within threadblock
    __syncthreads();
  }
// write result for this block to global mem
  if (tid == 0){
        opartNum_d[blockIdx.x] = idata1[0];
        opartDen_d[blockIdx.x] = idata2[0];
  }
}
```

3 Experiments

The datasets of n-grams available at [10] were considered for the experiments. They consist in histograms of 1-grams and 2-grams, representing text transcriptions extracted from 15-s video-clips—broadcast by CNN, Fox News and MSNBC—where either Mueller or Trump were mentioned. There are 218, 135 videos mentioning Mueller and 2, 226, 028 video clips that mentioned Trump. The datasets originally distinguished not just the news station but also the date of the emission, such that new histograms are computed each day. In our experiments, however, we only preserved the distinction of the media station but did not take into account the date of the recording; that is, the daily histograms were fused in our case. As a result, the histograms in our setup correspond to four problems, namely: histograms for 1-grams and histograms for 2-grams, either for Mueller or Trump separately. Finally, in order to allow the comparison of the histograms with the PBR distance in each problem, we expanded the local dictionaries of each news station to a global one that is common to the three of them. The details of the four problems, along with the sizes of the dictionaries, are shown in Table 1. Notice that the size of the histograms in the Trump's problems are, in both cases, almost 5 times grater than the Mueller's problems. All the experiments were carried out in a computer with the following specifications: Dell Computer, 64-bit architecture, with Intel® Xeon® CPU E5-2643 v3 @ 3.40 GHz CPU, Tesla K40c GPU and 64 GiB RAM.

Table 1. Problems and datasets of n-grams considered for the experiments.

(a) Problem 1: Mueller 1-grams

	Dictionary size	
Dataset	local	global
CNN: 1-Grams	31,421	59,717
Fox News: 1-Grams	25,615	
MSNBC: 1-Grams	40,701	

(b) Problem 2: Trump 1-grams

	Dictionary size	
Dataset	local	global
CNN: 1-Grams	138,304	282,000
Fox News: 1-Grams	129,720	
MSNBC: 1-Grams	155,017	

(c) Problem 3: Mueller 2-grams

	Dictionary size	
Dataset	local	global
CNN: 2-Grams	341,055	739,915
Fox News: 2-Grams	246,971	
MSNBC: 2-Grams	462,516	

(d) Problem 4: Trump 2-grams

	Dictionary size	
Dataset	local	global
CNN: 2-Grams	2,017,229	4,219,523
Fox News: 2-Grams	1,898,296	
MSNBC: 2-Grams	2,253,165	

Since each problem is composed by three histograms, its nine pairwise comparisons can be stored in a 3×3 distance matrix. Moreover, since the PBR distance fulfills the indiscernibility condition, we only used the values outside the main diagonal of the matrix for the sake of reporting the average and standard deviation of six computing performances, see Table 2. Elapsed times (ETs), in seconds, of the sequential version are reported in Fig. 1a and those of the parallel version in Fig. 1b. The corresponding speed-ups are also presented in Fig. 1c.

Table 2. Results of 6×25 executions for computing the PBR distance, cf. Fig. 1. Elapsed times are reported in seconds.

Problem	Mueller 1-grams	Trump 1-grams	Mueller 2-grams	Trump 2-grams
Mean ET CPU	0.0016 ± 0.0006	0.0085 ± 0.0036	0.0181 ± 0.0042	0.0954 ± 0.0045
Mean ET GPU	0.0001 ± 0.0597	0.0006 ± 0.0874	0.0010 ± 0.1151	0.0063 ± 0.3670
Speed-up	12.0229	15.5643	17.6008	15.1208

The ETs in CPU increase by a large percentage as the dataset sizes increase too. Although, the same happens with ETs in GPU, the growth is minimal compared to those in CPU, see Fig. 1. The standard deviations in GPU are significantly smaller than those of the executions in CPU; this may be explained by the fact that the CPU core must take care of not just the computations but also the administration of the scheduler of the processes. In contrast, the GPU threads are entirely dedicated to the computation task. The achieved accelerations with the GPU-based implementation ranged from 12 to 15 times with respect to the sequential version for the 1-gram problems and between 15 to 17 times for the 2-gram ones, see Table 2.

It can be seen that the performance improvement with the parallelized version is noticeable when using the benefits of many-core architectures. The speed-ups might become even more significant—in absolute terms—when considering collections with thousands or even millions of documents to be compared instead of the four problems presented in our experiments; among the paradigmatic examples of huge collection of documents, the Google Books repository and the Internet Archive are worth to be mentioned.

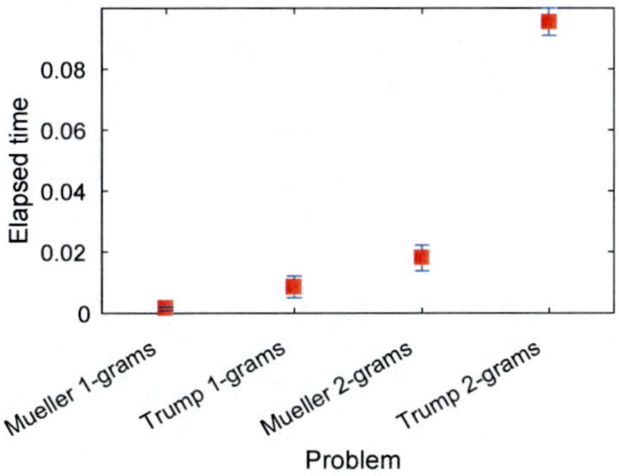

(a) Elapsed times in seconds: sequential version.

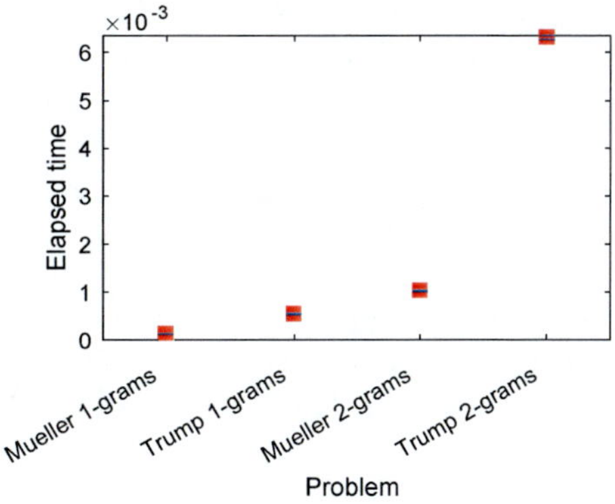

(b) Elapsed times in seconds: parallel version.

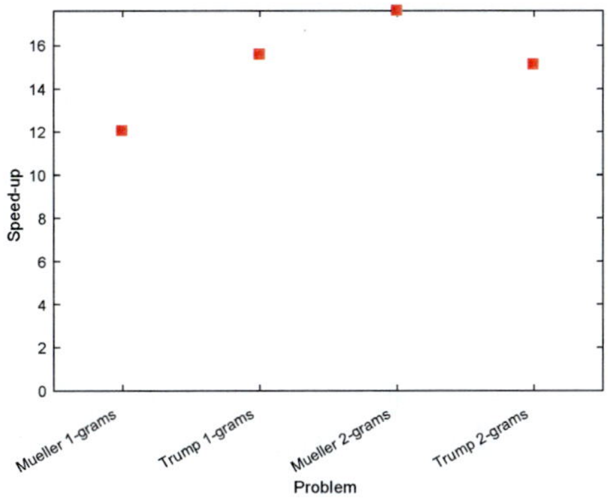

(c) Speed-ups.

Fig. 1. The two figures above show the sequential and parallel elapsed times. The figure on the left presents the speed-ups of the means calculated like of sequential over the parallel elapsed time $S_{up} = ET_{seq}/ET_{par}$.

4 Conclusion

This paper showed the computational benefit of parallelizing the computation of the PBR distance, particularly for comparing very long histograms (with lengths of up to 4 million bins) and making use of many-core architectures. Such long histograms are common in NLP applications, for instance in those based on bag-of-words representations. In order to reduce the sequential elapsed times of the execution of the PBR distance for histograms, we have proposed two kernel functions in GPU for adding vectors with a bin-to-bin approach and summing up the resulting vector via the reduction GPU strategy. In this contribution, the CUDA C codes of the kernel functions were provided and the results with four datasets of n-grams, exhibiting large histograms, were presented. It was shown that the proposed parallel implementation of the PBR distance reduces the computational complexity of the corresponding sequential algorithm, allowing to increase the speed-up of the sequential version up to 17 times with a large problem of 2-grams and running the algorithm on a Tesla 40c GPU. Future work includes testing the implementation with significantly larger problems as well as using more sophisticated versions of the parallel sum reduction.

Acknowledgments. The authors acknowledge support to attend DCAI'21 provided by Facultad de Administración and "Convocatoria nacional para el apoyo a la movilidad internacional 2019–2021", Universidad Nacional de Colombia Sede - Manizales.

References

1. Bicego, M., Londoño-Bonilla, J.M., Orozco-Alzate, M.: Volcano-seismic events classification using document classification strategies. In: Murino, V., Puppo, E. (eds.) ICIAP 2015. LNCS, vol. 9279, pp. 119–129. Springer, Cham (2015). https://doi.org/10.1007/978-3-319-23231-7_11
2. Bramer, M.: Text mining. In: Bramer, M.: Principles of Data Mining. Undergraduate Topics in Computer Science, 3rd edn, pp. 329–343. Springer, London (2016). https://doi.org/10.1007/978-1-4471-7307-6_20
3. Cheng, J., Grossman, M., McKercher, T.: Chapter 3: Cuda execution model. In: Cheng, J., Grossman, M., McKercher, T.: Professional CUDA C Programming, vol. 53, pp. 110–112. Wiley, Indianapolis (2013)
4. Ionescu, R.T., Popescu, M.: Object recognition with the bag of visual words model. Ionescu, R.T., Popescu, M.: Knowledge Transfer Between Computer Vision and Text Mining: Similarity-based Learning Approaches. ACVPR, pp. 99–132. Springer, Cham (2016). https://doi.org/10.1007/978-3-319-30367-3_5
5. Ishiguro, K., Yamada, T., Araki, S., Nakatani, T., Sawada, H.: Probabilistic speaker diarization with bag-of-words representations of speaker angle information. IEEE Trans. Audio Speech Lang. Process. **20**(2), 447–460 (2012). https://doi.org/10.1109/tasl.2011.2151858

6. Orozco-Alzate, M.: Recent (dis)similarity measures between histograms for recognizing many classes of plant leaves: an experimental comparison. In: Tibaduiza-Burgos, D.A., Anaya Vejar, M., Pozo, F. (eds.) Pattern Recognition Applications in Engineering, Advances in Computer and Electrical Engineering, chap. 8, pp. 180–203. IGI Global, Hershey (2020). https://doi.org/10.4018/978-1-7998-1839-7.ch008
7. Smith, S.W.: Chapter 2: Statistics, probability and noise. In: Smith, S.W.: Digital Signal Processing: A Practical Guide for Engineers and Scientists, pp. 11–34. Demystifying Technology. Newnes, Burlington (2002). https://doi.org/10.1016/b978-0-7506-7444-7/50039-x
8. Strubell, E., Ganesh, A., McCallum, A.: Energy and policy considerations for deep learning in NLP. In: Proceedings of the 57th Annual Meeting of the Association for Computational Linguistics, pp. 3645–3650. Association for Computational Linguistics, Florence (2019). https://doi.org/10.18653/v1/p19-1355
9. Swaminathan, M., Yadav, P.K., Piloto, O., Sjöblom, T., Cheong, I.: A new distance measure for non-identical data with application to image classification. Pattern Recogn. **63**, 384–396 (2017). https://doi.org/10.1016/j.patcog.2016.10.018
10. The GDELT Project: Two new ngram datasets for exploring how television news has covered Trump and Mueller (2019). https://tinyurl.com/242jswwb

Check for updates

CVAE-Based Complementary Story Generation Considering the Beginning and Ending

Riku Iikura(✉), Makoto Okada, and Naoki Mori

Osaka Prefecture University, 1-1 Gakuen-cho, Naka-ku, Sakai, Osaka, Japan
iikura@ss.cs.osakafu-u.ac.jp, {okada,mori}@cs.osakafu-u.ac.jp

Abstract. We studied the problem of the computer-based generation of a well-coherent story. In this study, we propose a model based on a conditioned variational autoencoder that takes the first and final sentences of the story as input and generates the story complementarily. One model concatenates sentences generated forward from the story's first sentence as well as sentences generated backward from the final sentence at appropriate positions. The other model also considers information of the final sentence in generating sentences forward from the first sentence of the story. To evaluate the generated story, we used the story coherence evaluation model based on the general-purpose language model newly developed for this study, instead of the conventional evaluation metric that compares the generated story with the ground truth. We show that the proposed method can generate a more coherent story.

1 Introduction

Automatic story generation is a frontier in the research area of neural language generation. In this study, we tackled the problem of automatically generating a well-coherent story by using a computer. In general, stories such as novels and movies are required to be coherent, that is, the story's beginning and end must be properly connected by multiple related events with emotional ups and downs. Against this background, we set up a new task to generate a story by giving the first and final sentences of the story as inputs and complementing them.

In this paper, we propose two models based on a conditioned variational autoencoder (CVAE) [1]. One model concatenates sentences generated forward from the first sentence of the story as well as sentences generated backward from the final sentence at appropriate positions, named *Story Generator Concatenating Two Stories* (SG-Concat). The other model also considers information of the final sentence in the process of generating sentences forward from the first sentence of the story, named *Story Generator Considering the Beginning and Ending* (SG-BE).

In the variational hierarchical recurrent encoder-decoder (VHRED) [1] and variational hierarchical conversation RNN (VHCR) [2], which are used for our models, higher quality dialogue generation has become possible by introducing a

© The Author(s), under exclusive license to Springer Nature Switzerland AG 2022
K. Matsui et al. (Eds.): DCAI 2021, LNNS 327, pp. 22–31, 2022.
https://doi.org/10.1007/978-3-030-86261-9_3

variational autoencoder (VAE) structure [3] for hierarchical recurrent encoder-decoder (HRED) [4]. HRED proposed for the dialogue-generation task has the advantage of being able to generate subsequent remarks while considering the contents of past remarks, so we expect that a well-coherent story can be generated. Although there have been studies focusing on story coherence [5,6], the approach of effectively using the information on the ending of stories has hardly been examined so far.

To evaluate the generated story, we used the story coherence evaluation model based on the general-purpose language model newly developed for this study, instead of the conventional evaluation metric that compares the generated story with the ground truth. Through an experiment, we showed that the proposed method can generate a coherent story. Our main contributions are summarized as follows:

- We propose a method to generate a story complementarily from the first and final sentences of the story as a new framework for story generation.
- We propose a new metric for evaluating the consistency of the story by using a language model that learned the task of detecting story breakdown.
- Our experimental results show that gradually increasing the influence of the story's final sentence make SG-BE generate more coherent and high-quality story.

2 Related Works

The number of studies on story generation has increased with the development of deep learning technology. Many previous studies used the sequence-to-sequence model (Seq2Seq), which records high accuracy in sentence generation tasks such as machine translation and sentence summarization. In theory, a recurrent neural network (RNN) learns to predict the next character or word in a sentence, as well as the probability of a sentence appearing. Roemmele et al. [7] tackled the Story Cloze task, which is a type of RNN that uses long short-term memory to generate the appropriate final sentence for a given context. Models based on Seq2Seq are typically trained to generate a single output. However, multiple endings are possible when considering the context of the story. In order to deal with this problem, Gupta et al. [8] proposed a method to generate various story endings by weighting important words in context and promoting the output of infrequently occurring words.

Several studies focus on the coherence of stories automatically generated by computers. For example, Fan et al. [5] proposed a hierarchical story generation system that combines the operation of generating plots to keep the story consistent and converting it into a story. Yao et al. [6] created a storyline from a given title or topic and created a story based on it to improve its quality. In addition, a method of generating a story from a subject with latent variables to learn the outline of the story [9] and a method of combining a story generation model and an explicit text planning model [10] were also proposed.

The main difference between these and our proposed approach is that not only is the first sentence of the story provided as input, but the final sentence is also provided. We propose a model based on CVAE proposed in the study of dialogue generation, which is a method of recursively generating sentences. We can expect this to generate a coherent story.

3 Technical Background

We consider a story as a sequence of N sentences $T = \{S_1, \ldots, S_N\}$. Each S_n contains a sequence of M_n tokens, that is, $S_n = \{w_{n,1}, \ldots, w_{n,M_n}\}$, in which $w_{n,m}$ is a random variable that takes values in the vocabulary V and represents the token at position m in sentence n. A generative model of the story parameterizes a probability distribution P_θ, which is controlled by the parameter θ, for any possible story. The probability of a story T can be decomposed as follows:

$$\begin{aligned} P_\theta(S_1, \ldots, S_N) &= \prod_{n=1}^{N} P_\theta(S_n | S_{<n}), \\ &= \prod_{n=1}^{N} \prod_{m=1}^{M_n} P_\theta(w_{n,m} | S_{<n}, w_{n,<m}), \end{aligned} \tag{1}$$

In this equation, $S_{<n} = \{S_1, \ldots, S_{n-1}\}$, $w_{n,<m} = \{w_{n,1}, \ldots, w_{n,m-1}\}$, that is, the tokens preceding m in the sentence S_n.

3.1 Hierarchical Recurrent Encoder Decoder

The HRED [4] is an extension of the encoder-decoder architecture in Seq2Seq for dialog generation. HRED assumes that each output sequence can be modeled into a two-level hierarchy: sequences of subsequences and subsequences of tokens. For example, a natural-language document can be modeled as a sequence of sentences (subsequences), with each sentence modeled as a sequence of words.

The HRED model consists of three RNN modules: an *encoder* RNN, a *context* RNN, and a *decoder* RNN. Each subsequence of tokens is deterministically encoded into a real-valued vector through the *encoder* RNN. This is given as input to the *context* RNN, which updates its internal hidden state to reflect all information up to that point in time. The *context* RNN deterministically outputs a real-valued vector, which the *decoder* RNN conditions on to generate the next subsequence of tokens.

3.2 Variational Hierarchical Recurrent Encoder Decoder

Serban et al. [1] proposed a variational hierarchical recurrent encoder-decoder (VHRED) by combining HRED with VAE [3], which introduced a latent variable to characterize the sentence-level representation for learning holistic properties.

VHRED uses a stochastic latent variable, $z_n \in \mathbb{R}^{d_z}$, for each subsequence $n = 1, \ldots, N$ conditioned on all previously observed tokens.

$$P_\theta(z_n|S_n) = \mathcal{N}(\mu(S_{<n}), \sum(S_{<n})). \tag{2}$$

In this equation, $\mathcal{N}(\mu, \sum)$ is the multivariate normal distribution with a mean of $\mu \in \mathbb{R}^{d_z}$ and a covariance matrix of $\sum \in \mathbb{R}^{d_z \times d_z}$. In other words, the latent variable z_n is a value sampled with probability distribution following the mean and variance of the input sentence representation up to the current time step, and the probability distribution that follows the variance, and generating randomness in the output sentence is possible. Given z_n, the model generates the n-th subsequence tokens $S_n = (w_{n,1}, \ldots, w_{n,M_n})$ as follows:

$$P_\theta(S_n|z_n, S_{<n}) = \prod_{m=1}^{M_n} P_\theta(w_{n,m}|z_n, S_{<n}, w_{n,<m}). \tag{3}$$

3.3 Variational Hierarchical Conversation RNN

The Variational Hierarchical Conversation RNN (VHCR) [2] model involves two key ideas: using hierarchical structure of latent variables and exploiting an utterance drop regularization. A global latent variable is introduced to generate a sequence of utterances in a dialog. The main difference from VHRED is that, in addition to the local latent variables for each sentence, the global latent variables for each conversation are used to form a hierarchical latent structure. This global latent variable follows $\mathcal{N}(0, I)$ and is invariant in one conversation.

4 Complementary Story Generation

4.1 Story Generator Concatenating Two Stories

The SG-Concat model concatenates a story generated from the first sentence of the story and a story generated in the reverse order from the final sentence at appropriate positions. Specifically, this model follows this procedure:

1. Prepare two models that have been trained to generate stories forward and backward. Further in this paper, each model is referred to as the *forward model* and the *backward model* for convenience. Each model are based on VHRED or VHCR.
2. Concatenate the stories generated by the *forward model* and the *backward model* at the appropriate positions.

Forward model $\mathcal{G}^{\mathrm{F}}$ and *backward model* $\mathcal{G}^{\mathrm{B}}$ take the sequence of a sentence as an input and then output the following sentence:

$$\mathcal{G}^{\mathrm{F}}(S_1, \ldots, \hat{S}^{\mathrm{F}}_{n-1}) = \hat{S}^{\mathrm{F}}_n, \tag{4}$$

$$\mathcal{G}^{\mathrm{B}}(S_N, \ldots, \hat{S}^{\mathrm{B}}_{n+1}) = \hat{S}^{\mathrm{B}}_n, \tag{5}$$

Here, $\hat{S}$ is a sentence generated by the model. The position of sentence n is based on data used by the *forward model* for training. SG-Concat concatenates stories generated by the *forward model* and the *backward model* at the sentence position i $(2 \leq i \leq N-1)$ that maximizes the value of a certain evaluation function $\mathcal{M}$. In this paper, we used BERTScore [11] as the evaluation function $\mathcal{M}$:

$$i = \arg\max_{n} \mathcal{M}(\hat{S}_n^{\mathrm{F}}, \hat{S}_n^{\mathrm{B}}). \tag{6}$$

Finally, the story $\hat{T}$ generated by SG-Concat is a sequence of sentences, as follows:

$$\hat{T} = \begin{cases} \{S_1, \hat{S}_2^{\mathrm{F}}, \ldots, \hat{S}_{i-1}^{\mathrm{F}}, \hat{S}_i^{\mathrm{F}}, \hat{S}_{i+1}^{\mathrm{B}}, \ldots, \hat{S}_{N-1}^{\mathrm{B}}, S_N\} & (\text{if } i \leq \lceil \frac{N}{2} \rceil) \\ \{S_1, \hat{S}_2^{\mathrm{F}}, \ldots, \hat{S}_{i-1}^{\mathrm{F}}, \hat{S}_i^{\mathrm{B}}, \hat{S}_{i+1}^{\mathrm{B}}, \ldots, \hat{S}_{N-1}^{\mathrm{B}}, S_N\}. & (\text{if } i > \lceil \frac{N}{2} \rceil) \end{cases} \tag{7}$$

4.2 Story Generator Considering the Beginning and Ending

The SG-BE model takes the first and final sentences of a story as input and generates one story to complement the sentences between them. Figure 1 shows an overview of SG-BE based on VHRED. The output sentence is predicted as follows:

$$P_\theta(S_n | S_{<n}, S_N) = \prod_{m=1}^{M_n} P_\theta(w_{n,m} | S_{<n}, w_{n,<m}, S_N). \tag{8}$$

In the original VHRED and VHCR, which are based on HRED, the final hidden state of the *encoder* RNN for the sentences generated in order from the input sentence is treated as the input to the *context* RNN. However, in the input to the *context* RNN in SG-BE, the information of the distributed representation of the final sentence of the story is also considered. The input to the *context* RNN at time step t $(1, \ldots, N-1)$ for that purpose was determined by using the following two methods:

No weight the final sentence: The sum of the encoder hidden state h_{S_t} of the sentence S_t and the encoder hidden state h_{S_N} of the sentence S_N that is, $h_{S_t} + h_{S_N}$ is used as an input of the *context* RNN. In this case, the influence of the final sentence S_N at each time step is equal.

Weight the final sentence: The value $h_{S_t} + \frac{t}{N-1} h_{S_N}$ calculated from the encoder hidden state h_{S_t} of the sentence S_t and the encoder hidden state h_{S_N} of the sentence S_N is used as an input of the *context* RNN. In this case, the influence of the final sentence S_N becomes stronger as the time step passes.

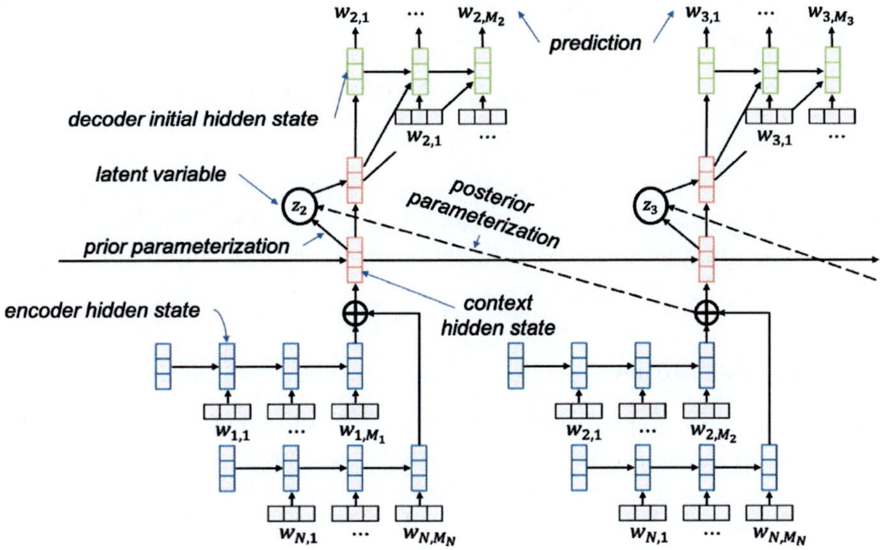

Fig. 1. The overview of SG-BE based on VHRED. Each sentence is encoded by *encoder* RNN, mapped to the context of the story, and then used to generate the tokens in the next sentence. In the process of sequentially generating sentences from the first sentence of the story, a hidden state of *encoder* RNN for the final sentence is added. The input format for SG-BE based on VHCR is the same.

5 Evaluation Experiment

5.1 Dataset

We used the ROCStories [12] corpus to generate the stories in this experiment. The sample contained in this dataset is a story, each consisting of five sentences. The inputs for the *forward model* and the *backward model* make up the first sentences of the story and the outputs for them make up the last four sentences of the story. The dataset for the *backward model* is the reverse order of the sentences of each story in the dataset for the *forward model.* However, the inputs for SG-Concat and SG-BE are the first and final sentences, and their outputs are the second, third, and fourth sentences of the story.

We used the Python-based natural language toolkit NLTK [13] to perform tokenization and named-entity recognition. All names were replaced with the `<person>` and converted to lowercase. Words with a frequency of occurrence of three or more were registered as vocabularies in the training data, and the number of vocabularies was 16,700. There were 13,818 words that occurred less than three times, and they were replaced by `<unk>`. We randomly divided the 98,161 stories contained in the corpus into 78,528:9,816:9,817 (= 80:10:10), each of which was used for training, validation, and testing.

5.2 Hyper-parameters

Each model was constructed based on the two sentence generation models: VHRED [1], and VHCR [2]. The parameters common to all models were set to the same values based on the research by Park et al. [2] The embedding size of the word was 500, and a two-layer gated recurrent unit was adopted for each RNN unit. We applied a dropout ratio of 0.2 during training. The batch size was 64. For optimization, we used Adam with a learning rate of 0.0001 with gradient clipping.

5.3 Evaluation Metrics

Because the purpose of this study is to generate a new coherent story, we do not use an evaluation metric that refers to a single sentence as a ground-truth like BLEU. The evaluation metrics used, Distinct-N and *story coherence score*, are described as follows:

Distinct-N: Distinct-1,2,3 calculates numbers of distinct unigrams, bigrams, and trigrams in the generated responses divided by the total numbers of unigrams, bigrams, and trigrams. Higher Distinct scores indicated higher diversity in the generated outputs.

Story coherence score: In this study, we propose a *story coherence score* (SCS) as a metric for automatically evaluating the coherency of generated sentences. This was inspired by the evaluation metric based on the model that learned the Story Cloze task, which was proposed by Gupta et al. [8] We used a task to identify a randomly rearranged story consisting of five sentences (negative example) and ground-truth (positive example). To solve this problem, we fine-tuned RoBERTa [14] implemented by huggingface (https://github.com/huggingface/transformers), which is called the *Story Coherence Model* (SCM). For the winner, we pick the story with a greater probability (from RoBERTa's output head) of being the ground truth. The numbers of positive and negative examples in the task to be learned were the same. By using this approach, we could obtain an accuracy of 95.9%. We defined the ratio of examples identified as the ground truth in the dataset as the SCS.

5.4 Results and Analysis

Table 1 shows examples of stories. Example 1 is generated by SG-BE (No weight the final sentence) based on VHCR, and example 2 was generated by SG-BE (Weight the final sentence) based on VHCR. As a result of estimation by SCM, example 1 was presumed to be a negative example, and example 2 was presumed to be a positive example. It seems that the fourth sentence of example 2 is semantically more strongly linked to the final sentence of the story than example 1. Furthermore, the story generated in SG-BE (Weight the final sentence) (Example 3) was identified as a positive example, while the ground-truth

Table 1. Examples of stories. The boldfaced sentences are sentences in the actual story.

Example 1 (SG-BE (No weight the final sentence) based on VHCR):
<person> was going to ride a bike for the first time. He did n't know how to play it. It did n't take long to get better. He went to a professional. **It made him give up trying.**
Example 2 (SG-BE (Weight the final sentence) based on VHCR):
<person> was going to ride a bike for the first time. He tried to ride with his friends. He tried to get off the ride. Unfortunately it started to rain. **It made him give up trying.**
Example 3 (Ground-truth):
betty always brought her lunch to work. One day, she left her lunch at home and was very mad. She decided she would not buy lunch out. She was so hungry all afternoon. That night, she put a reminder on her phone about lunch.
Example 4 (SG-BE (Weight the final sentence) based on VHCR):
Betty always brought her lunch to work. One day she forgot her lunch. She had no choice but to call her mom to help her. Her mom was upset that she did n't know what to do. **That night, she put a reminder on her phone about lunch.**

Table 2. Test set results for each model. "Base Model" refers to the model used when constructing each story generation model. The best performance is boldfaced.

Base model	Distinct-1	Distinct-2	Distinct-3	SCS
Ground-truth	0.0278	0.243	0.584	97.1%
Baseline: *Forward model*				
VHRED	0.0188	0.147	0.353	81.8%
VHCR	0.0186	0.151	0.374	84.1%
Baseline: *Backward model*				
VHRED	0.0190	0.150	0.364	79.2%
VHCR	0.0185	0.147	0.361	79.7%
Our model: SG-Concat				
VHRED	0.0190	0.148	0.360	78.5%
VHCR	0.0186	0.149	0.369	79.7%
Our model: SG-BE (No weight the final sentence)				
VHRED	0.0186	0.155	0.393	84.0%
VHCR	0.0185	**0.158**	**0.410**	84.3%
Our model: SG-BE (Weight the final sentence)				
VHRED	0.0189	0.151	0.372	83.2%
VHCR	0.0187	0.157	0.404	**85.5%**

(Example 4) was identified as a negative example. Thus, it was observed that the proposed model was able to generate a high-quality story.

Table 2 presents the evaluation results. The evaluation values for the original sample (Ground-truth) in the ROCStories corpus are also shown as reference values. We confirmed that the evaluation value of the story generated by VHCR regardless of the input to the *context* RNN, excluding Distinct-1, is relatively high. From this result, it can be said that the story becomes lexically diverse and more coherent by introducing global latent variables.

By comparing the *forward model* with the *backward model*, notice that the story generated by the *forward model* has a higher SCS. SCS for the new story—generated by concatenating the stories generated by these two models at appropriate positions by SG-Concat was lower than that for each story before the concatenation. From this, coherency is considered lost when the stories generated by different models are concatenated.

We consider that the SCS of the story generated by SG-BE is higher than that of the story generated by any other model. Specifically, the SCS is higher when the final sentence is weighted. Thus, it is considered that a more coherent story can be generated by gradually increasing the influence of the final sentence in the process of sequentially generating sentences from the first sentence.

6 Conclusion

In this paper, we propose a method to generate a story complementarily from the first and final sentences of the story as a new framework for story generation. Through the evaluation experiment, we examined a method to effectively add the information of the final sentence to the input of the *context* RNN of the hierarchical encoder decoder model in which the latent variable is introduced. The quality of the generated story was evaluated from the viewpoint of consistency by using SCM, which learned the task of detecting broken narrative sentences by randomly rearranging sentences. The experiment confirmed that the model that concatenates the story generated from forward and the story generated from backward cannot generate a coherent story. We also confirmed that a more coherent and high-quality story can be generated by gradually increasing the influence of the final sentence in the process of sequentially generating sentences from the first sentence of the story.

In the evaluation experiment, we defined a new automatic quantitative evaluation metric. Our future work will include the following: (1) evaluation of the qualitative quality of the generated story, and (2) improvement of the model to generate longer and more coherent stories.

Acknowledgments. This work was supported by JSPS KAKENHI Grant, Grant-in-Aid for Scientific Research(B), 19H04184. This work was also supported by JSPS KAKENHI Grant, Grant-in-Aid for Scientific Research(C), 20K11958.

References

1. Serban, I.V., et al.: A hierarchical latent variable encoder-decoder model for generating dialogues. In: Proceedings of the Thirty-First AAAI Conference on Artificial Intelligence, pp.3295–3301 (2017)
2. Park, Y., Cho, J., Kim, G.: A hierarchical latent structure for variational conversation modeling. In: Proceedings of the 2018 Conference of the North American Chapter of the Association for Computational Linguistics: Human Language Technologies, pp.1792–1801 (2018)
3. Kingma, D.P., Welling, M.: Auto-encoding variational Bayes. In: 2nd International Conference on Learning Representations (2014)
4. Serban, I.V., Sordoni, A., Bengio, Y., Courville, A., Pineau, J.: Building end-to-end dialogue systems using generative hierarchical neural network models. In: Proceedings of the Thirtieth AAAI Conference on Artificial Intelligence, pp.3776–3783 (2016)
5. Fan, A., Lewis, M., Dauphin, Y.: Hierarchical neural story generation. In: Proceedings of the 56th Annual Meeting of the Association for Computational Linguistics, pp.889–898 (2018)
6. Yao, L., Peng, N., Weischedel, R., Knight, K., Zhao, D., Yan, R.: Plan-and-write: towards better automatic storytelling. In: Proceedings of the AAAI Conference on Artificial Intelligence, pp.7378–7385 (2019)
7. Roemmele, M., Kobayashi, S., Inoue, N., Gordon, A.: An RNN-based binary classifier for the story cloze test. In: Proceedings of the 2nd Workshop on Linking Models of Lexical, Sentential and Discourse-Level Semantics, pp.74–80 (2017)
8. Gupta, P., Kumar, V.B., Bhutani, M., Black, A.W.: WriterForcing: generating more interesting story endings. In: Proceedings of the Second Workshop on Storytelling, pp. 117–126 (2019)
9. Chen, G., Liu, Y., Luan, H., Zhang, M., Liu, Q., Sun, M.: Learning to predict explainable plots for neural story generation. arXiv preprint arXiv:1912.02395 (2019)
10. Zhai, F., Demberg, V., Shkadzko, P., Shi, W., Sayeed, A.: A hybrid model for globally coherent story generation. In: Proceedings of the Second Workshop on Storytelling of the Association for Computational Linguistics, pp. 34–45 (2019)
11. Zhang, T., Kishore, V., Wu, F., Weinberger, K.Q., Artzi, Y.: BERTScore: evaluating text generation with BERT. In: International Conference on Learning Representations (2020)
12. Mostafazadeh, N., et al.: A corpus and cloze evaluation for deeper understanding of commonsense stories. In: Proceedings of the 2016 Conference of the North American Chapter of the Association for Computational Linguistics: Human Language Technologies, pp. 839–849 (2016)
13. Bird, S., Loper, E., Klein, E.: Natural Language Processing with Python. O'Reilly Media Inc., Newton (2009)
14. Liu, Y., et al.: RoBERTa: a robustly optimized BERT pretraining approach. arXiv preprint arXiv:1907.11692 (2019)

Check for updates

A Review on Multi-agent Systems and Virtual Reality

Alejandra Ospina-Bohórquez[1](✉), Sara Rodríguez-González[2], and Diego Vergara-Rodríguez[3]

[1] Department of Computer Science and Automation, Faculty of Science, University of Salamanca, Plaza de los Caídos s/n, 37008 Salamanca, Spain
ale.ospina15@usal.es
[2] University of Salamanca, BISITE, Digital Innovation Hub, Calle Espejo s/n, 37007 Salamanca, Spain
srg@usal.es
[3] Department of Mechanical Engineering, Catholic University of Ávila, C/Canteros, s/n, 05005 Avila, Spain
diego.vergara@ucavila.es

Abstract. The objective of this work is to perform a *Systematic Mapping Study* (SMS) on works that deal with the combined use of *Multi-agent Systems* (MAS) and *Virtual Reality* (VR). 83 relevant studies were found, which were categorized according to their purpose. This review led to establishing current and future lines of research on the subject.

Keywords: Multi-agent Systems · Virtual Reality · Systematic Mapping Study

1 Introduction

A Systematic Mapping Study (SMS) was done according to the methodology of Kietchenham et al. [1] and Petersen et al. [2]. It allowed a categorization in the areas of Multi-agent Systems (MAS) and Virtual Reality (VR). A SMS permits establishing the frequency in which investigations are performed on a specific subject. At first, VR was used mostly for military purposes, however, with the appearance of game engines (tools for the development of virtual environments), applications in different fields have begun to be developed Now, VR has become a suitable technique for visualization, simulation, design, etc. in different areas such as videogames, education, architecture, etc. It allows the user to interact with an environment that may not be accessible in the real world. A MAS is a system that can be decomposed into entities called agents. An agent is an autonomous entity capable of communicating, taking action, and interact with other agents. In MAS, groups of agents act together (cooperating or competing, sharing or not sharing knowledge, etc.) to achieve the systems' goals. The combination of MAS and VR has allowed the development of more interactive and realistic applications. It is intended to show the scope of research in this area.

© The Author(s), under exclusive license to Springer Nature Switzerland AG 2022
K. Matsui et al. (Eds.): DCAI 2021, LNNS 327, pp. 32–42, 2022.
https://doi.org/10.1007/978-3-030-86261-9_4

2 Research Methodology

Based on the methodology proposed by Kitchenham et al. [1] and Peterson et al. [2], a SMS including three phases was done: Planning, Development and Report.

Table 1. Evaluation realized in this SMS according to the guide of Petersen et al. [2]

Research process			Done
Planning	Need for the map	1. Motivate the need and relevance	Yes
		2. Define objective and research questions	Yes
		3. Consult with target audience to define questions	No
Development	Choosing search strategy	4. Snowballing	No
		5. Manual search	No
		6. Database search	Yes
	Develop the search	7. PICO	Yes
		8. Consult experts	No
		9. Iteratively improve search	Yes
		10. Keyword of known papers	Yes
		11. Use standards, encyclopedias	No
	Search evaluation	12. Paper test-set	Yes
		13. Expert evaluation	No
		14. Authors' web pages	No
		15. Test-retest	No
	Inclusion/Exclusion criteria	16. Identify objective criteria for decision	Yes
		17. Resolve disagreements among multiple researchers	Yes
		18. Decision rules	Yes
Report	Extraction process	19. Identify objective criteria for decision	Yes
		20. Obscuring information that could bias	No
		21. Resolve disagreements among multiple researchers	Yes
		22. Test-retest	No
	Classification scheme	23. Research type	Yes
		24. Research method	Yes
		25. Venue type	Yes
	Validity and discussion	26. Validity discussion/limitations provided	Yes

2.1 Planning

Motivation and Objective: This study aims to establish the current situation and future research lines of the combined use of VR and MAS.

Research Questions: These questions will allow to categorize the studies *(i) What applications have been developed combining VR and MAS? (ii) What benefits does the combined use of these technologies bring?*

2.2 Development of the Study

Search Strategy: The *Population*, *Intervention*, *Comparison* and *Outcomes* (PICO) methodology proposed by Kitchenham et al. [1] was chosen as the search strategy. **(i) Population:** research papers, **(ii) Intervention:** multi-agent architectures used in VR applications, **(iii) Comparison:** analogy of the studies, **(iv) Outcomes:** n/a.

Inclusion Criteria: (i) Studies about MAS and VR. (ii) Studies from 2016 to 19 April 2021.

Exclusion Criteria: (i) Duplicated articles. (ii) Articles where the authors doubt about the contribution. (iii) Articles that are not in English or Spanish.

2.3 Mapping Report

Filtering Studies: 613 studies were found then the first author applied the inclusion/exclusion criteria, in case of doubt the other authors were asked.

Classification Process: The studies were categorized according to the keywords and titles of the articles, in line with the purpose of the works: videogames, robots, human behaviour, autonomous vehicles and machinery simulation, intelligent virtual environments (IVEs), cultural heritage and urban development.

Validation Process: Petersen et al. [2] indicated the activities that should be done within a SMS. A well-done SMS must perform at least 33% of the activities established. In Table 1 it is shown the activities performed (61%).

3 Mapping

The database search returned 613 studies. The inclusion/exclusion criteria were applied. The filtering process is illustrated in Fig. 1a and b shows the categorization done.

4 Discussion

In the area of machinery simulation, MAS are used for parts validation and personnel training [3–5]. For example, in [3] a MAS is proposed to establish a route planner for a manufacturing line and its effectiveness is verified through a

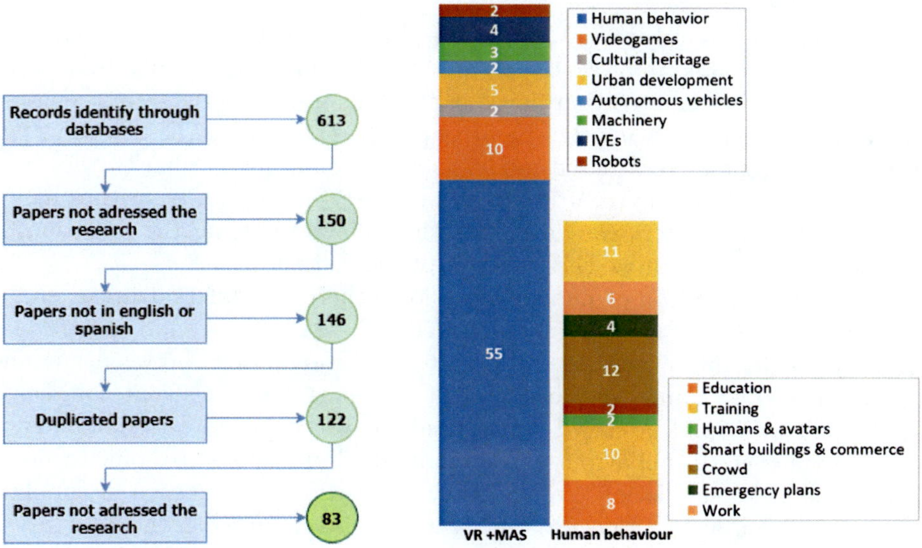

(a) Filtering process

(b) Categorization of studies

Fig. 1. Study process and results

simulation in VR. In [4], the authors propose a VR simulator with MAS-based methods for the joint operation of the machinery of a coal mine. And in [5], a virtual multi-agent-based maintenance system is established where humans in training, tools and equipment are considered agents.

In autonomous vehicle simulation, MAS are used for the validation of vehicles in the real world [6,7], e.g. In [7], a simulation platform is proposed that assesses the perception, planning algorithms and decision-making of vehicles, using a multi-agent interaction system.

MAS are also used to improve urban planning [8–12], e.g. In [9], a system for the simulation and visualization of pedestrians was established that includes a MAS to assign physical and psychological characteristics to the virtual agents.

MAS and VR also served for the dissemination of cultural heritage [13,14], e.g. in [13], a recreation of the agora of Athens is made in a virtual world, in which the MAS simulates the human behaviour of the time.

In the field of human behaviour MAS are used to create movements and behaviours [15–25]. For example, [15] establishes an algorithm to generate realistic movements for human agents and integrates it with Unreal Engine to demonstrate the benefits it brings. In [18] is presented a motivated learning agent that has drives that establish its motivations respond to events and create goals.

In the area of robot simulation, agents are used for their representation and validation [26,27]. For example, [26] presents a solution based on imitation, to learn collaborative strategies between multiple agents. Its usefulness was evaluated in a simulation environment of the RoboCup Soccer3D.

In the case of video games, MAS are used mainly to create characters, generate better stories and improve the quality of the games [28–37], e.g. In [31], an intelligent agent was created that develops its skills from experience in the game and makes decisions following the cognitive patterns of humans. In [29], a system is proposed using an agent-based model to create and improve backstories in a way that promotes the myth of the hero's journey. In [30], an algorithm based on multi-agent simulation was developed that allows inferring the intention of the user's avatar, to improve the response of the virtual agents.

Within the simulation of human behaviour, there are different areas: simulation of crowds [38–49], emergency plans [50–53], work [54–58] and educational environments [59–66], personnel training [67–76], interactions between humans and avatars [77,78], and the development of smart buildings and commerce [79,80]. E.g. In [38,39], a crowd simulation was proposed, multi-agent models are used to change the behaviour of each individual in response to some external event. In [52], the simulation of emergency evacuations was proposed, simulating emotional contagion in groups during events. [54] focuses on the integration in workspaces of disabled people, improving the accessibility of buildings and adapting work processes, the authors proposed a MAS with the ability to perform social simulations in 3D environments. In [63], a human behaviour simulator was presented that uses a MAS to emulate a virtual audience so that it can be used for teacher training. In [72], an environment for teaching medicine was presented where medical cases are simulated to evaluate the knowledge of the students, a MAS is used for the classification of virtual patients. In [77], an approach is proposed to generate realistic interactions between virtual agents and user avatars in complex multi-agent and multi-avatar environments. In [79], a multi-agent simulation tool of a mall was established, it allows determining the behaviour of agents within the system.

5 Results

5.1 What Applications Have Been Developed Combining VR and MAS?

Different applications were found with diverse purposes that were used to make the categorization: machinery [3–5], human behaviour [15–25], and robot simulation [26,27], autonomous vehicles [6,7], urban [8–12], and videogames development [28–37] and dissemination of cultural heritage [13,14]. And within the simulation of human behaviour, subcategories were established: crowds [38–49], emergency plans [50–53], and work [54–58] and educational environment simulations [59–66], personnel training [67–76], interactions humans-avatars [77,78], and smart buildings/commerce [79,80].

5.2 What Benefits Does the Combined Use of These Technologies Bring?

MAS allow much more efficient development of VR applications. In machinery, autonomous vehicle, and robot simulations, MAS are used to simulate objects for their validation and staff training [3–7,26,27]. In urban development, MAS serve to simulate objects to improve urban planning [9–12]. For cultural heritage dissemination, MAS are used to simulate objects, people and cultures [13,14]. In human behaviour simulation, MAS are used to create movements [15,16] and behaviours [17–25] for characters. In videogames, MAS are used to create characters [31–34], to generate better stories [29] and improving quality [28,30].

6 Conclusions

Research questions were set: *What apps have been developed combining VR and MAS?* and *What benefits does the combined use VR and MAS bring?* that permitted to establish the research scope and the studies categorization. The use of MAS and VR allowed the development of higher-quality applications in the areas of machinery, human behaviour and robot simulation, autonomous vehicles, IVEs, urban and videogames development, and cultural heritage dissemination.

Acknowledgements. This research has been supported by the project "Intelligent and sustainable mobility supported by multi-agent systems and edge computing (InEDGEMobility): Towards Sustainable Intelligent Mobility: Blockchain-based framework for IoT Security", Reference: RTI2018-095390-B-C32, financed by the Spanish Ministry of Science, Innovation and Universities (MCIU), the State Research Agency (AEI) and the European Regional Development Fund (FEDER).

References

1. Kitchenham, B., Budgen, D., Brereton, P.: Using mapping studies as the basis for further research – participant-observer case study. Inf. Softw. Technol. **53**, 638–651 (2011)
2. Petersen, K., Vakkalanka, S., Kuzniarz, L.: Guidelines for conducting systematic mapping studies in software engineering: an update. Inf. Softw. Technol. **64**, 1–18 (2015)
3. Durica, L., Gregor, M., Vavrík, V., Marschall, M., Grznár, P., Mozol, Š.: A route planner using a delegate multi-agent system for a modular manufacturing line: proof of concept. Appl. Sci. **9**, 4515 (2019)
4. Xie, J., Yang, Z., Wang, X., Zeng, Q., Li, J., Li, B.: A virtual reality collaborative planning simulator and its method for three machines in a fully mechanized coal mining face. Arab. J. Sci. Eng. **43**, 4835–4854 (2018)
5. Wang, Y., Lv, C., Zhou, D., Yu, D., Peng, X.: Multi-agent based modeling and simulation of virtual maintenance system. In: Proceedings of WCICA 2016, pp. 2963–2968. IEEE, New York (2016)
6. Elmquist, A., Hatch, D., Serban, R., Negrut, D.: An overview of a connected autonomous vehicle emulator (CAVE). In: Proceedings of IDETC/CIE 2017, pp. 1–12. ASME, New York (2017)
7. Chen, Y., Chen, S., Zhang, T., Zhang, S., Zheng, N.: Autonomous vehicle testing and validation platform: integrated simulation system with hardware in the loop*. In: Proceedings of IV 2018, pp. 949–956. IEEE, New York (2018)

8. Ren, J., Xiang, W., Xiao, Y., Yang, R., Manocha, D., Jin, X.: Heter-sim: heterogeneous multi-agent systems simulation by interactive data-driven optimization. IEEE Trans. Vis. Comput. Graph. **27**, 1953–1966 (2019)
9. Rivalcoba, I., Toledo, L., Rudomín, I.: Towards urban crowd visualization. Sci. Vis. **11**, 39–55 (2019)
10. Okamoto, S., Takematsu, S., Matsumoto, S., Otabe, T., Tanaka, T., Tokuyasu, T.: Development of design support system of a lane for cyclists and pedestrians. In: Proceedings of CISIS 2016, pp. 385–388. IEEE, New York (2016)
11. Chen, A.Y., Chen, J.H.: Urban rail transit operation simulation based on virtual reality technology. In: Proceedings of CICTP 2017, pp. 1736–1745. ASCE, Reston (2017)
12. Garg, D., Chli, M., Vogiatzis, G.: Traffic3D: a new traffic simulation paradigm. In: Proceedings of AAMAS 2019, pp. 2354–2356. International Foundation for Autonomous Agents and Multiagent Systems, Richland (2019)
13. Vosinakis, S., Avradinis, N., Koutsabasis, P.: Dissemination of intangible cultural heritage using a multi-agent virtual world. In: Ioannides, M., Martins, J., Žarnić, R., Lim, V. (eds.) Advances in Digital Cultural Heritage. LNCS, vol. 10754, pp. 197–207. Springer, Cham (2018). https://doi.org/10.1007/978-3-319-75789-6_14
14. Kiourt, C., Pavlidis, G., Koutsoudis, A., Kalles, D.: Multi-agents based virtual environments for cultural heritage. In: Proceedings of ICAT 2017, pp. 1–6. IEEE, New York (2017)
15. Narang, S., Best, A., Manocha, D.: Simulating movement interactions between avatars & agents in virtual worlds using human motion constraints. In: Proceedings of IEEE VR 2018, pp. 9–16. IEEE, New York (2018)
16. Cafaro, A., Ravenet, B., Ochs, M., Vilhjálmsson, H.H., Pelachaud, C.: The effects of interpersonal attitude of a group of agents on user's presence and proxemics behavior. ACM Trans. Interact. Intell. Syst. **6**, 12 (2016)
17. Song, Y., Niu, L., Li, Y.: Individual behavior simulation based on grid object and agent model. ISPRS Int. J. Geo-Inf. **8**, 388 (2019)
18. Starzyk, J.A., Graham, J., Puzio, L.: Needs, pains, and motivations in autonomous agents. IEEE Trans. Neural Netw. Learn. Syst. **28**, 2528–2540 (2017)
19. Bönsch, A., Vierjahn, T., Shapiro, A., Kuhlen, T.: Turning anonymous members of a multiagent system into individuals. In: Proceedings of IEEE VHCIE, pp. 1–4. IEEE, New York (2017)
20. Zhang, X., Schaumann, D., Haworth, B., Faloutsos, P., Kapadia, M.: Coupling agent motivations and spatial behaviors for authoring multiagent narratives. Comput. Animat. Virtual Worlds **30**, e1898 (2019)
21. Puig, X., et al.: VirtualHome: simulating household activities via programs. In: Proceedings of IEEE/CVF CVPR, pp. 8494–8502. IEEE, New York (2018)
22. Andelfinger, P., et al.: Incremental calibration of seat selection preferences in agent-based simulations of public transport scenarios. In: Proceedings of WSC, pp. 833–844. IEEE, New York (2018)
23. Bera, A., Randhavane, T., Kubin, E., Shaik, H., Gray, K., Manocha, D.: Data-driven modeling of group entitativity in virtual environments. In: Proceedings VRST 2018, pp. 1–10. Association for Computing Machinery, New York (2018)
24. Ranjbartabar, H., Richards, D.: A virtual emotional freedom therapy practitioner: (demonstration). In: Proceedings of AAMAS 2016, pp. 1471–1473. International Foundation for Autonomous Agents and Multiagent Systems, Richland (2016)
25. Ohmoto, Y., Marimoto, T., Nishida, T.: Effects of the perspectives that influenced on the human mental stance in the multiple-to-multiple human-agent interaction. Procedia Comput. Sci. **112**, 1506–1515 (2017)

26. Raza, S., Haider, S.: Using imitation to build collaborative agents. ACM Trans. Auton. Adapt. Syst. **11**, 3 (2016)
27. Seghour, S., Tadjine, M.: Consensus-based approach and reactive fuzzy navigation for multiple no-holonomic mobile robots. In: Proceedings of ICSC 2017, pp. 492–497. IEEE, New York (2017)
28. Lakshika, E., Barlow, M., Easton, A.: Understanding the interplay of model complexity and fidelity in multi-agent systems via an evolutionary framework. IEEE Trans. Comput. Intell. AI Games **9**, 277–289 (2017)
29. García-Ortega, R., García-Sánchez, P., Merelo Guervós, J., San-Ginés, A., Fernández-Cabezas, A.: The story of their lives: massive procedural generation of Heroes' journeys using evolved agent-based models and logical reasoning. In: Squillero, G., Burelli, P. (eds.) EvoApplications 2016. LNCS, vol. 9597, pp. 604–619. Springer, Cham (2016). https://doi.org/10.1007/978-3-319-31204-0_39
30. Narang, S., Best, A., Manocha, D.: Inferring user intent using Bayesian theory of mind in shared avatar-agent virtual environments. IEEE Trans. Vis. Comput. Graph. **25**, 2113–2122 (2019)
31. Makarov, I., Tokmakov, M., Poluakov, P.: First-person shooter game for virtual reality headset with advanced multi-agent intelligent system. In: Proceedings of MM 2016, pp. 735–736. Association for Computing Machinery, New York (2016)
32. Seele, S., Haubrich, T., Schild, J., Herpers, R., Grzegorzek, M.: Augmenting cognitive processes and behavior of intelligent virtual agents by modeling synthetic perception. In: Proceedings of the MM 2017, pp. 117–125. Association for Computing Machinery, New York (2017)
33. Seele, S., Haubrich, T., Schild, J., Herpers, R., Grzegorzek, M.: Integration of multi-modal cues in synthetic attention processes to drive virtual agent behavior. In: Beskow J., Peters, C., Castellano, G., O'Sullivan, C., Leite, I., Kopp, S. (eds.) IVA 2017. LNCS, vol. 10498, pp. 403–412. Springer, Cham (2017). https://doi.org/10.1007/978-3-319-67401-8_50
34. Nunnari, F., Héloir, A.: Yet another low-level agent handler. Comput. Animat. Virtual Worlds **30**, e1891 (2019)
35. Matthews, J., Charles, F., Porteous, J., Mendes, A.: MISER: Mise-En-ScèNe region support for staging narrative actions in interactive storytelling. In: Proceedings of AAMAS 2017, pp. 782–790. International Foundation for Autonomous Agents and Multiagent Systems, Richland (2017)
36. Matthews, J., Charles, F., Porteous, J., Mendes, A.: Mise-En-ScèNe of narrative action in interactive storytelling. In: Proceedings of AAMAS 2017, pp. 1799–1801. International Foundation for Autonomous Agents and Multiagent Systems, Richland (2017)
37. Porteous, J., Charles, F., Smith, C., Cavazza, M., Mouw, J., van den Broek, P.: Using virtual narratives to explore children's story understanding. In: Proceedings of AAMAS 2017, pp. 773–781. International Foundation for Autonomous Agents and Multiagent Systems, Richland (2019)
38. Kim, S., Bera, A., Best, A., Chabra, R., Manocha, D.: Interactive and adaptive data-driven crowd simulation. In: Proceedings of 2016 IEEE VR, pp. 29–38. IEEE, New York (2016)
39. Bera, A., Kim, S., Manocha, D.: Interactive and adaptive data-driven crowd simulation: user study. In: Proceedings of 2016 IEEE VR, p. 325. IEEE, New York (2016)

40. Phon-Amnuaisuk, S., Rafi, A., Au, T.W., Omar, S., Voon, N.H.: Crowd simulation in 3D virtual environments. In: Sombattheera, C., Stolzenburg, F., Lin, F., Nayak, A. (eds.) MIWAI 2016. LNCS, vol. 10053, pp. 162–172. Springer, Cham (2016). https://doi.org/10.1007/978-3-319-49397-8_14
41. Wang, X., et al.: Crowd formation via hierarchical planning. In: Proceedings of VRCAI 2016, pp. 251–260. Association for Computing Machinery, New York (2016)
42. Agıl, U., Güdükbay, U.: A group-based approach for gaze behavior of virtual crowds incorporating personalities. Comput. Animat. Virtual Worlds **29**, e1806 (2018)
43. Narang, S., Best, A., Randhavane, T., Shapiro, A., Manocha, D.: PedVR: simulating gaze-based interactions between a real user and virtual crowds. In: Proceedings of VRST 2016, pp. 91–100. Association for Computing Machinery, New York (2016)
44. Novick, D., Hinojos, L.J., Rodriguez, A.E., Camacho, A., Afravi, M.: The market scene: physical interaction with multiple agents. In: Proceedings of HAI 2018, pp. 387–388. Association for Computing Machinery, New York (2018)
45. Randhavane, T., Bera, A., Manocha, D.: F2Fcrowds: planning agent movements to enable face-to-face interactions. Presence Teleop. Virtual Environ. **26**, 228–246 (2017)
46. Dickinson, P., Gerling, K., Hicks, K., Murray, J., Shearer, J., Greenwood, J.: Virtual reality crowd simulation: effects of agent density on user experience and behaviour. Virtual Real. **23**, 19–32 (2019)
47. Montana, L., Maddock, S.: A sketch-based interface for real-time control of crowd simulations that use navigation meshes. In: Proceedings of the VISIGRAPP 2019, pp. 41–52. SciTePress, Setúbal (2018)
48. Jayalath, C., Wimalaratne, P., Karunananda, A.: Modelling goal selection of characters in primary groups in crowd simulations. Int. J. Simul. Model. **15**, 585–596 (2016)
49. Chen, H., Wong, S.K.: Transporting objects by multiagent cooperation in crowd simulation: transporting objects by multi-agent cooperation. Comput. Animat. Virtual Worlds **29**, e1826 (2018)
50. Li, Y., Hu, B., Zhang, D., Gong, J., Song, Y., Sun, J.: Flood evacuation simulations using cellular automata and multiagent systems - a human-environment relationship perspective. Int. J. Geogr. Inf. Sci. **33**, 2241–2258 (2019)
51. Wang, Y., Wang, L., Liu, J.: Object behavior simulation based on behavior tree and multi-agent model. In: Proceedings of 2017 IEEE ITNEC, pp. 833–836. IEEE, New York (2017)
52. Mao, Y., Yang, S., Li, Z.: Personality trait and group emotion contagion based crowd simulation for emergency evacuation. Multimed. Tools Appl. **79**, 3077–3104 (2020)
53. Montecchiari, G., Bulian, G., Gallina, P.: Towards real-time human participation in virtual evacuation through a validated simulation tool. J. Risk Reliab. **232**, 476–490 (2018)
54. Barriuso, A., De La Prieta, F., Villarrubia, G., Hernández de la Iglesia, D., Lozano Murciego, Á.: MOVICLOUD: agent-based 3D platform for the labor integration of disabled people. Appl. Sci. **8**, 337 (2018)
55. Zeng, Y., Zhang, Z., Han, T.A., Spears, I.R., Qin, S.: Using intention recognition in a simulation platform to assess physical activity levels of an office building. In: Proceedings of AAMAS 2017, pp. 1817–1819. International Foundation for Autonomous Agents and Multiagent Systems, Richland (2017)
56. Antakli, A., et al.: Agent-based web supported simulation of human-robot collaboration. In: Proceedings of the WEBIST 2019, pp. 88–99. SciTePress, Setúbal (2019)

57. Antakli, A., Zinnikus, I., Klusch, M.: ASP-driven BDI-planning agents in virtual 3D environments. In: Klusch, M., Unland, R., Shehory, O., Pokahr, A., Ahrndt, S. (eds.) MATES 2016. LNCS, vol. 9872, pp. 198–214. Springer, Cham (2016). https://doi.org/10.1007/978-3-319-45889-2_15
58. Cai, L., Liu, B., Yu, J., Zhang, J.: Human behaviors modeling in multi-agent virtual environment. Multimed. Tools Appl. **76**, 5851–5871 (2017)
59. Calvo, O., Molina, J., Patricio, M.A., Berlanga, A.: A propose architecture for situated multi-agent systems and virtual simulated environments applied to educational immersive experiences. In: Ferrández Vicente, J., Álvarez-Sánchez, J., de la Paz López, F., Toledo Moreo, J., Adeli, H. (eds.) IWINAC 2017. LNCS, vol. 10338, pp. 413–423. Springer, Cham (2017). https://doi.org/10.1007/978-3-319-59773-7_42
60. Baierle, I.L.F., Gluz, J.C.: Programming intelligent embodied pedagogical agents to teach the beginnings of industrial revolution. In: Nkambou, R., Azevedo, R., Vassileva, J. (eds.) ITS 2018. LNCS, vol. 10858, pp. 3–12. Springer, Cham (2018). https://doi.org/10.1007/978-3-319-91464-0_1
61. Tazouti, Y., Boulaknadel, S., Fakhri, Y.: ImALeG: a serious game for Amazigh language learning. Int. J. Emerg. Technol. Learn. (iJET) **14**, 28–38 (2019)
62. Boulaknadel, S., Tazouti, Y., Fakhri, Y.: Towards a serious game for Amazigh language learning. In: Proceedings of 2019 IEEE/ACS 16th AICCSA, pp. 1–5. IEEE, New York (2019)
63. Nilsson, J., Klügl, F.: Human-in-the-loop simulation of a virtual classroom. In: Rovatsos, M., Vouros, G., Julian, V. (eds.) EUMAS 2015, AT 2015. LNCS, vol. 9571, pp. 379–394. Springer, Cham (2016). https://doi.org/10.1007/978-3-319-33509-4_30
64. Lugrin, J.L., et al.: Benchmark framework for virtual students' behaviours. In: Proceedings of AAMAS 2018, pp. 2236–2238. International Foundation for Autonomous Agents and Multiagent Systems, Richland (2018)
65. Barange, M., Saunier, J., Pauchet, A.: Pedagogical agents as team members: impact of proactive and pedagogical behavior on the user. In: Proceedings of AAMAS 2017, pp. 791–800. International Foundation for Autonomous Agents and Multiagent Systems, Richland (2017)
66. Fukuda, M., Huang, H.H., Nishida, T.: Investigation of class atmosphere cognition in a VR classroom. In: Proceedings of 6th HAI 2018, pp. 374–376. Association for Computing Machinery, New York (2018)
67. Blankendaal, R.A., Bosse, T.: Using run-time biofeedback during virtual agent-based aggression de-escalation training. In: Demazeau, Y., An, B., Bajo, J., Fernández-Caballero, A. (eds.) PAAMS 2018. LNCS, vol. 10978, pp. 97–109. Springer, Cham (2018). https://doi.org/10.1007/978-3-319-94580-4_8
68. Feng, D., Jeong, D., Krämer, N., Miller, L., Marsella, S.: Is it just me?: evaluating attribution of negative feedback as a function of virtual instructor's gender and proxemics. In: Proceedings of AAMAS 2017, pp. 810–818. International Foundation for Autonomous Agents and Multiagent Systems, Richland (2017)
69. Johnson, E., Gratch, J., DeVault, D.: Towards an autonomous agent that provides automated feedback on students' negotiation skills. In: Proceedings of AAMAS 2017, pp. 410–418. International Foundation for Autonomous Agents and Multiagent Systems, Richland (2017)
70. Tavcar, A., Gams, M.: Surrogate-agent modeling for improved training. Eng. Appl. Artif. Intell. **74**, 280–293 (2018)

71. Barthès, J.P.A., Wanderley, G.M.P., Lacaze-Labadie, R., Lourdeaux, D.: Designing training virtual environments supported by cognitive agents. In: Proceedings of 2018 IEEE CSCWD, pp. 295–300. IEEE, New York (2018)
72. De Lima, R.M., et al.: A 3D serious game for medical students training in clinical cases. In: Proceedings of 2016 IEEE SeGAH, pp. 1–9. IEEE, New York (2016)
73. Benkhedda, S., Bendella, F.: FASim: a 3D serious game for the first aid emergency. Simul. Gaming. **50**, 690–710 (2019)
74. Ooi, S., Tanimoto, T., Sano, M.: Virtual reality fire disaster training system for improving disaster awareness. In: Proceedings of ICEIT 2019, pp. 301–307. Association for Computing Machinery, New York (2019)
75. Tianwu, Y., Changjiu, Z., Jiayao, S.: Virtual reality based independent travel training system for children with intellectual disability. In: Proceedings of UKSim-AMSS 2016, pp. 143–148. IEEE, New York (2016)
76. Sánchez San Blas, H., Sales Mendes, A., García Encinas, F., Silva, L.A., González, G.V.: A multi-agent system for data fusion techniques applied to the internet of things enabling physical rehabilitation monitoring. Appl. Sci. **11**, 331 (2021)
77. Best, A., Narang, S., Manocha, D.: SPA: verbal interactions between agents and avatars in shared virtual environments using propositional planning. In: Proceedings of 2020 IEEE VR, pp. 117–126. IEEE, New York (2020)
78. Braz, P., Werneck, V.M.B., de Souza Cunha, H., da Costa, R.M.E.M.: SMEC-3D: a multi-agent 3D game to cognitive stimulation. In: Bajo, J., et al. (eds.) PAAMS 2018. CCIS, vol. 887, pp. 247–258. Springer, Cham (2018). https://doi.org/10.1007/978-3-319-94779-2_22
79. Christian, J., Hansun, S.: Simulating shopper behavior using fuzzy logic in shopping center simulation. J. ICT Res. Appl. **10**, 277–295 (2016)
80. Zhao, Y., Pour, F., Golestan, S., Stroulia, E.: BIM Sim/3D: multi-agent human activity simulation in indoor spaces. In: Proceedings of 2019 IEEE/ACM 5th International Workshop on SEsCPS, pp. 18–24. IEEE, New York (2019)

Malware Analysis with Artificial Intelligence and a Particular Attention on Results Interpretability

Benjamin Marais[1,2(✉)], Tony Quertier[1], and Christophe Chesneau[2]

[1] Orange Labs, Lannion, France
{benjamin.marais,tony.quertier}@orange.com
[2] Department of Mathematics, LMNO, University of Caen Normandy, Caen, France
christophe.chesneau@unicaen.fr

Abstract. Malware detection and analysis are active research subjects in cybersecurity over the last years. Indeed, the development of obfuscation techniques, as packing, for example, requires special attention to detect recent variants of malware. The usual detection methods do not necessarily provide tools to interpret the results. Therefore, we propose a model based on the transformation of binary files into grayscale image, which achieves an accuracy rate of 88%. Furthermore, the proposed model can determine if a sample is packed or encrypted with a precision of 85%. It allows us to analyze results and act appropriately. Also, by applying attention mechanisms on detection models, we have the possibility to identify which part of the files looks suspicious. This kind of tool should be very useful for data analysts, it compensates for the lack of interpretability of the common detection models, and it can help to understand why some malicious files are undetected.

1 Introduction

In recent years, the number of malware and attacks has increased exponentially. The illustration of this phenomenon is the number of online submissions to sandboxes, such as Virustotal or Any.run, among other things. In addition, these malware are increasingly difficult to detect due to very elaborate evasion techniques. For example, polymorphism is used to evade pattern-matching detection relied on by security solutions like antivirus software, while some characteristics of polymorphic malware change, its functional purpose remains the same. These developments make obsolete detection solutions like signature-based detection. Researchers and companies have therefore turned to artificial intelligence methods to manage both large volumes and complex malware. In this paper, we will look at the static analysis of malware for computational issues such as time and resources. Indeed, dynamic analysis gives very good results, but for companies that have thousands of suspicious files to process, it creates resource problems because a sandbox can require two to three minutes per file.

© The Author(s), under exclusive license to Springer Nature Switzerland AG 2022
K. Matsui et al. (Eds.): DCAI 2021, LNNS 327, pp. 43–55, 2022.
https://doi.org/10.1007/978-3-030-86261-9_5

1.1 State of Art

Malware detection and analysis represent very active fields of study. In recent years, several methods have been proposed in this regard.

The most popular detection method is signature-based detection [1,2]. This method consists of stocking portions of code of benign and malicious files called signatures. It consists of comparing the signature of a suspicious file with the signature database. A weakness of this method is having the file first, determining its nature and recorded its signature.

Another common and effective method is called dynamic analysis. It attempts to run suspect files in secure environments (physical or virtual) named sandbox [3]. It allows analysts to study the behavior of the file without risk. This process is particularly effective in detecting new malware or known malware that has been modified with obfuscation techniques. This procedure, however, may be a waste of time and resources. Also, some malware is able to detect virtual environments and does not run to hide its nature and behavior.

In order to achieve good results in malware detection, and overcome signature-based detection and dynamic analysis weaknesses, many approaches to static analysis associated with machine learning have been investigated in recent works. Static analysis aims to study a file without running it to understand its purpose and nature. The most natural way is to extract features based on binary file bit statistics (entropy, distributions...) then to use ML algorithms to perform a binary classification (Random Forest, XGBoost, LightGBM for example). Among other things, the quality of detection models depends on features used for training and on the amount of data. In this regard, Anderson et al. [4] provide Ember, a very good dataset to train ML algorithms. On the other hand, Raff et al. [5] use Natural Language Processing tools to analyse bit sequences extracted from binary files. Their MalConv algorithm gives very good results but requires a lot of computing power to train it. Moreover, it has recently been shown that this technique is very vulnerable to padding and GAN-based evasion methods. To overcome these weaknesses, Fleshman et al. [6] developed Non-Negative MalConv which reduces the evasion rate but provides a slight drop in accuracy.

Nataraj et al. [7] introduced the use of grayscale images to classify 25 malware families. Authors convert binary files into images and use the GIST algorithm to extract important features from them. They train a K-NN with these features and obtain a percentage of accuracy of 97.25%. In addition to presenting a good classification rate, this method has the characteristic of offering better resilience against obfuscation, especially against packing, the most used obfuscation method. In the continuity of this work, Vu et al. [8] proposed the use of RGB (for Red Green Blue) images for malware classification with their own transformation method called Hybrid Image Transformation (HIT). They encode the syntactic information in the green channel of an RGB image, while the red and blue channels capture the entropy information.

In view of the interest in image recognition, with ImageNet [9] for example, and performance improvements [10] on the topic for several years, some authors proposed using a Convolutional Neural Network (CNN) apply to binary files converted into grayscale images for malware classification. Rezende [11] applied transfer learning on ResNet-50 for malware family classification and achieved a percentage of accuracy of 98.62%. To go deeper into the subject, Yakura et al. [12] used attention mechanism with CNN to highlight areas in grayscale images that help with classification. Also, they relate areas of importance to the disassembled function of the code.

Another principal trend in malware research is to protect detection models against obfuscation techniques. Many malware are known, but they have been modified to make them undetectable. For example, polymorphic [13] and metamorphic [14] malicious files embed mechanisms that modify their code apparently but not their behavior. Moreover, malware authors can alter them manually. Kreuk et al. [15] inject bytes directly into the binary file to perturb the detection model without transforming its functions. Another modification is to pack malware, and it is one of the commonly used methods to easily evade antivirus software. Aghakhani et al. [16] give an overview of the limits of detection models to spot packed malware.

1.2 Contributions and Paper Plan

The contributions of the study can be summarized as follows:

- Different detection methods are tested on a real database containing complex malware harvested in the company. In particular, we propose detection models that use grayscale image and HIT preprocessing on our own dataset of binary files. We compare the results of our models with models (LGBM, XGBoost, DNN) trained with the Ember dataset and preprocessing.
- We propose models that take into consideration the fact that binary files can be packed or encrypted. One objective of this method is to reduce the false positive rate due to the interpretation of some models that modified files are necessarily malicious. Another objective is to give malware analysts a tool that provides them with more information on the nature of a suspicious file.
- We implement attention mechanisms to interpret the results of our image recognition algorithms. This method is used to extract the parts of the image, and therefore of the binary, which contributed the most to the scoring of the classification. This allows this information to be passed on to security analysts to facilitate and accelerate reverse engineering work on the malware. Their feedback is then used to understand algorithm errors and improve this aspect.

This paper is organized as follows: in Sect. 2, we give a description of our dataset, discussing its advantages and the different preprocessing methods involved. In Sect. 3, we present the different models trained on Ember or on our own datasets. We compare the models and discuss the results and performances. Section 4 is dedicated to the analysis of modified samples and to attention mechanisms, two methods which can be an interesting aid for analysts. Section 5 summarizes the results and concludes the paper.

2 Dataset and Preprocessing

2.1 Description of Binaries Dataset

Our dataset contains 22,835 benign and malware in Portable Executable (PE) format, including packed or encrypted binary files. Figure 1 shows the exact distribution of the dataset. The benign files are derived from harvested Windows executables, and the malwares have been collected in companies and on sandboxes. The dataset's main distinguishing feature is that these malware are relatively difficult to detect. As evidence, they have been undetected by some sandboxes or antivirus programs. As our dataset contains complex and non-generic malware, it should prevent overfitting during the training of our models.

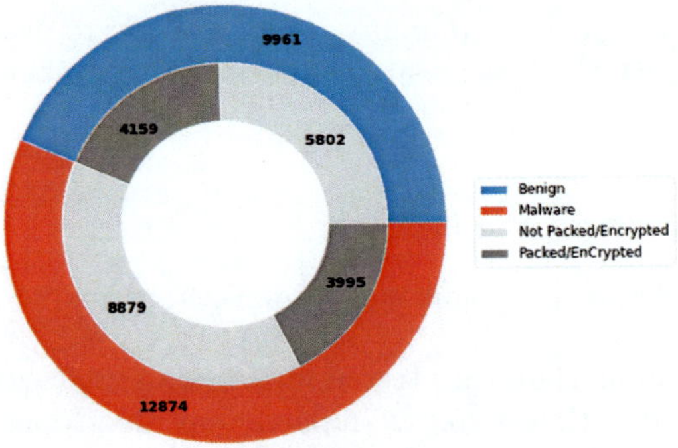

Fig. 1. Distribution of our dataset

To train machine learning algorithms, we use the Ember dataset which contains 600,000 PE files for training and we test on our own dataset to see the results. For the image-based algorithm, we split the dataset into 80% of training data, 10% of testing data and 10% of validation data. This distribution is the most optimized to keep a training sample large enough and a testing sample complex enough.

2.2 Is the Malware Modified?

A recurring problem when doing static analysis is the analysis of packed or encrypted executables, that we include under the term "modified" file in the rest of the paper. Artificial intelligence algorithms will often classify them as malicious even though many benign executables are modified for industrial or intellectual property reasons, for example. This is understandable given that these processes will drastically alter the entropy and distribution of bytes in the executable. A line of thought for better performance is to take into consideration the modifying nature of binary files during the training of detection models.

Upstream of the analysis, the use of software such as ByteHist [17] gives an idea of the nature of a file. Indeed, ByteHist is a tool for generating byte-usage-histograms for all types of files with a special focus on binary executables in PE-format. ByteHist allows us to see the distribution of bytes in an executable. The more the executable is packed, the more uniform the distribution is. Figure 2 presents some byte distribution examples of one malware and one benign not packed and their UPX-transformed equivalents.

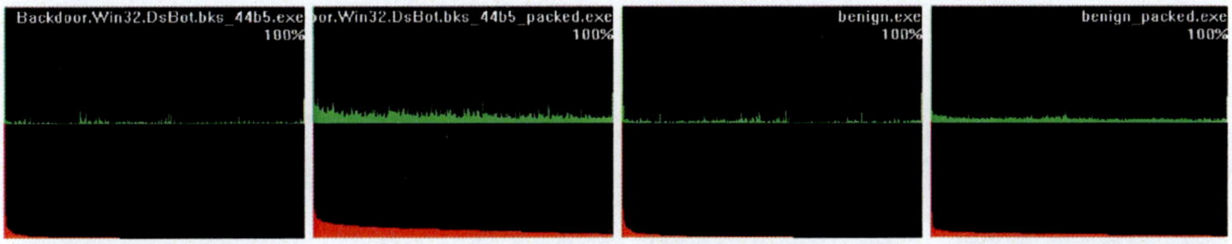

(a) Malware not packed (b) Malware packed (c) Benign not packed (d) Benign packed

Fig. 2. Byte distribution comparison between malware and benign with ByteHist

As we can see, UPX changes the byte distribution of binary files, in particular for the malware examples with more modifications than the benign file. Also, it is a common packer and it is easy to unpack binary files created with UPX. However, many malware are packed with more complex software, making analysis more difficult.

2.3 Image-Based Malware Transformation

Before discussing how to turn a binary into an image, let us briefly explain why we use images. First of all, different sections of a binary can be easily seen when it is transformed into an image, so that it can give a first orientation to an analyst as to where to look, as we will see in the next section. Then, as we discussed in the introduction, malware authors can modify parts of their files or use polymorphism to change their signatures or produce recent variants. Images can capture small alterations yet retain the global structure of the malware.

Given a static binary, we map it directly to an array of integers between 0 and 255. Hence, each binary is converted into a one-dimensional array $v \in [0, 255]$, v is then reshaped into a two-dimensional array and we follow the resizing scheme as presented in [7]. That is, the width is determined with respect to the file size. The height of the file is the total length of the one-dimensional array divided by the width. We rounded up the height and pad zeros if the width is undivisible by the file size. This method allows us to transform a binary into a grayscale image. The main advantage of this process is that it is very fast. For 20,000 binaries, it takes at most a few minutes.

Vu et al. [8] give different methods to transform a binary into an RGB image. Their color encoding scheme is based on the fact that green is the most sensitive to human vision and has the highest coefficient value in image grayscale conversion. In particular, with their HIT method, they encode the syntactic information into the green channel of an RGB image, while the red and blue channels capture the entropy information. In this way, clean files will intuitively have more green pixels than malicious files, which contain higher entropy with higher red/blue values. This transformation gives very good results with image recognition algorithms. The only downside is the transformation time. It takes an average of 25 s to transform a binary into an image with their HIT method.

Figure 3 presents grayscale and HIT transformations of the binary file introduced previously.

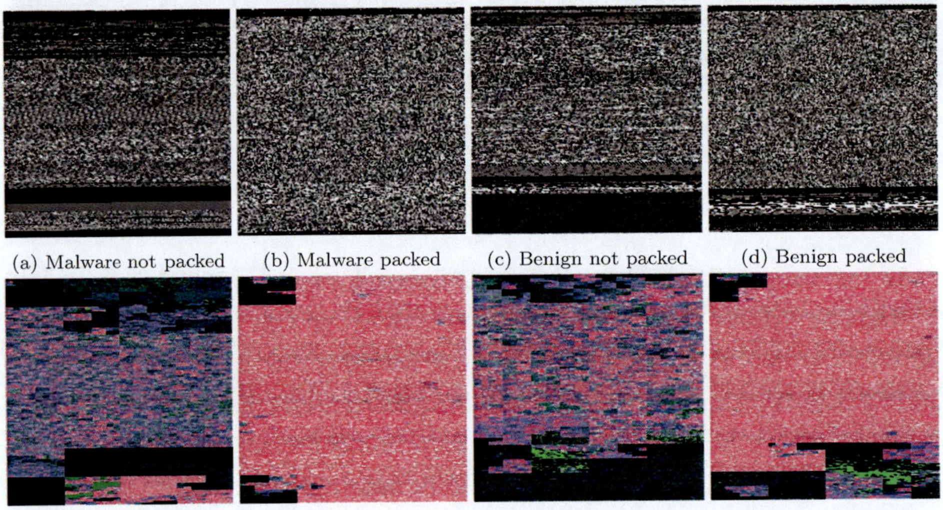

(a) Malware not packed (b) Malware packed (c) Benign not packed (d) Benign packed

(e) Malware not packed (f) Malware packed (g) Benign not packed (h) Benign packed

Fig. 3. Grayscale representation (top) and HIT representation (bottom) of some binary files

3 Detection Based on Static Methods

In this part, we study and compare three approaches to malware detection based on static methods and machine learning algorithms:

- First, we train three models on the Ember dataset with their own feature extraction method.
- Then, using this time grayscale images as input, we propose a CNN to detect malware and, to go further, three hybrid models.
- Finally, we train another CNN on an RGB image using the HIT method.

3.1 Algorithms on Binary Files

For the static analysis, we will test three algorithms: XGBoost, LightGBM and a deep neural network (DNN). XGBoost [18] is a reference algorithm for testing data but, on a large dataset, there can be some issues with computing time. That's why we also compare it with LightBGM [19] which is used by Ember in connection with their dataset.

Let us quickly introduce the LightGBM algorithm which is still less known. It uses a novel technique of Gradient-based One-Side Sampling (GOSS) to filter out the data instances to find a split value while XGBoost uses a pre-sorted algorithm and a histogram-based algorithm for computing the best split. Here, instances are observations. Its main advantages compared to other algorithms like Random Forest or XGBoost are:

- Faster training speed and higher efficiency.
- Lower memory usage (replaces continuous values with discrete bins, which results in lower memory usage).
- Better accuracy with more complex tree.

Specifically, if we focus on it in this study, it is mainly because of its capacity to handle a huge amount of data. It can perform equally well with large datasets and present a significant reduction in training time as compared to XGBoost.

To begin with, we train algorithms XGBoost and LightGBM on the Ember dataset, and we test them on our own data. In addition, we train a DNN on the Ember learning dataset too, because this kind of models goes hand to hand with a large dataset that contains so many features. We use the F1 score and accuracy score to compare models between them. Results are collected in Table 1.

From this table, we can see that LightGBM and DNN performances are very close, but that XGBoost is less performant (either in precision or computing time).

3.2 Algorithms on Grayscale Images

Based on Nataraj et al. [7] work, we transform our dataset into grayscale images and employ them to train CNN. Our CNN is composed of three convolutional layers, a dense layer with a ReLU activation function and a sigmoid function for scoring binaries. Also, inspired by [20], we propose hybrid models combining CNN and LightGBM, RF or Support Vector Machine (SVM). Firstly, we use CNN to reduce the number of dimensions and, for each binary image, we go from 4,096 features to 256. Then, we use these 256 new features to train RF, LightGBM and SVM models. As shown in Table 1, F1 and accuracy scores are still used to compare models.

As can be seen, the hybrid model combining CNN and RF outperforms the four grayscale models, but the overall results are close. Also, the performances are relatively close to those of the LightGBM and the DNN presented in Sect. 3.1. It should be noted that the grayscale models are trained using only 19,400 binary files, whereas the previous models' training set consists of 600,000 binary files. So, with the grayscale transformation and a dataset thirty times smaller, our grayscale models remain reliable for malware detection compared to conventional models and preprocessing.

3.3 Algorithms on RGB Images

We are now evaluating our CNN on the basis of RGB images using HIT transformation. Table 1 shows F1 and accuracy scores on the test sample.

Even if the performance of the RGB model is better than the others previously presented, training on a local machine is quite long with RGB images, but scoring on a single one is fast. Due to the complexity of HIT algorithm, the transformation time of binaries into images is quite long and takes, on average, 25 s for a sample, against less than one second for the grayscale transformation. First of all, it increases the learning process considerably if we include the time to convert the 24,000 samples. Moreover, when predicting malware, the score is obtained in less than one second, but the time for converting the binary into an image is added. Considering this, it's useless to use HIT transformation compared to grayscale transformation in corporate case, this is why we will not dwell on training other models with HIT preprocessing.

Table 1. Models F1 score and accuracy for Ember (left), grayscale (middle) and HIT (right)

	F1 score	Accuracy score
LGBM	0.9110	0.9001
XGBoost	0.8275	0.7748
DNN	**0.9160**	**0.9071**

	F1 score	Accuracy score
CNN	0.8786	0.8703
CNN+LGBM	0.8827	0.8703
CNN+RF	**0.8914**	**0.8804**
CNN+SVM	0.8895	0.8791

	F1 score	Accuracy score
CNN	0.934	0.94

4 Modified Binary Analysis and Attention Mechanism

The objective, in addition to having the most accurate results possible, is to make them usable by an analyst. To do this, we must be able to understand why our algorithms give high scores or not. This allows us to improve the learning process in case of error, but also to give a pointer to the analysts to know where to look. We propose two approaches to facilitate the understanding and analysis of malware:

- The first approach is to use, during the training of our algorithms, information about the nature of binary files. In particular, we know if the binary files of the training set are modified. The purpose is to reduce false positive results caused by these two obfuscation methods, and also give more information about the nature of the new suspect files.

- The second approach is the use of an attention mechanism on model trains with grayscale images. We can generate heatmap using the attention mechanism to detect suspicious patterns. Furthermore, the heatmap also helps to understand the results of the malware detection model.

4.1 Modified Binaries

In order to reduce the false positive rate due to obfuscation, we also provide two models which are trained while taking into account the altered nature of the binary file. Grayscale images are used as input for both models.

1. The first model is a CNN which returns output information on the nature of the binary file, if it is a malware or not, and if it is obfuscated or not. So, with a single CNN, we have double knowledge on the characteristics of the binary file. This model achieves a F1 score of 0.8924 and an accuracy score of 0.8852.
2. The second model is a superposition of three CNNs. The first one is used to separate binary files, according to whether they are obfuscated or not, with an accuracy of 85%. The two others are used to predict if a binary file is a malware or a benign and each model is respectively trained on modified binary file and not modified binary file. The main advantage of this model is that each CNN is independent from the other two and can be retrained separately. They also use different architectures to improve the generalization of the data used to train them. We get a F1 score of 0.8797 and an accuracy score of 0.8699 for this model.

As we can see, the first model gives better results than the second model. Also, it can determine if a binary file is modified with an accuracy rate of 84%. This information could help malware analysts to have a better expertise. For example, this can explain why some benign files are detected as malware. Moreover, it can encourage the use of sandboxes for certain suspicious files if they modified and if the result of the malware detection is ambiguous.

4.2 Interpretibility of Results and Most Important Bytes

In this section, we present an approach that can help analysts to interpret the results of detection models based on the transformation of binary files into grayscale images. A grayscale image representation of an executable has distinct texture depending on the relevant sections of the file [21]. We can use tools like PE file to extract information from the binary and see the relationship between the PE file and its grayscale image representation. Figure 4 shows an executable transformed into an image, the corresponding sections of the PE file (left), and information associated with each texture (right).

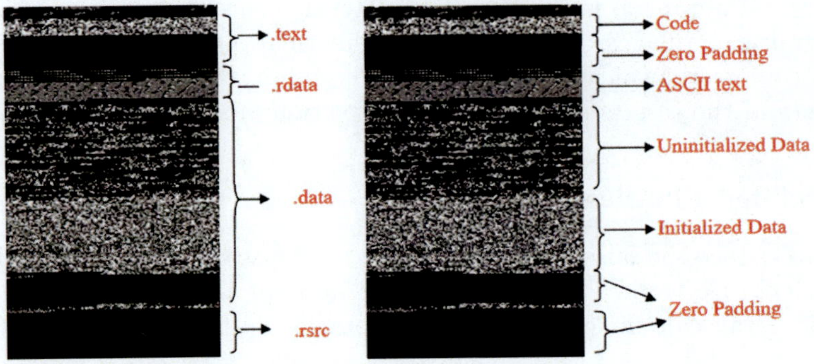

Fig. 4. Grayscale image with corresponding section (left) and texture interpretation (right)

The associations of PE file and grayscale image can allow an analyst to quickly visualize the areas of interest of the binary file. To go further in the analysis, it should be necessary to study which parts of the image have contributed to the results of malware detection algorithm. To do this, we use attention mechanisms that consist of highlighting the pixels that influence the most the prediction score of our algorithm.

We use GradCAM++ [22] with our own CNN presented in Sect. 3.2. The GradeCAM++ algorithm extracts from the CNN the pixels that influence the most the model's decision, i.e. those that determine if the file is benign or malicious. It returns heatmap which could be interpreted as follows, the more the image area impacts the CNN prediction, the warmer the coloring. Figure 5 presents heatmaps of the four binaries introduced in Sect. 2.2. We observe that the malware and its packed version do not have the same activation zones. We can make the same remark about the benign. Also, as the byte distribution of the packed malware has undergone a lot of modifications, we see that more zones and pixels are lit up compared to the original malware. This means that the model needs more information to make a decision.

On the other hand, this kind of representation could help to understand why a binary file is misclassified. For example, a common evasion technique is called padding and relies on adding byte sequences to artificially increase the size of a file and fool the antivirus. With image representation, this kind of technique is easily detected. However, as we can see in Fig. 6, for the two examples, the padding zone is considered as an important part of the file. Even if the malware is correctly detected, the benign file is misclassified and labeled as malware. So, the padding is considered as a malicious criterion. This knowledge could be taken into consideration to increase the performance of the malware detection model.

The activation map on binary file images seems to be an interesting and useful tool to help malware analysts. However, it is necessary to deepen this subject to fully exploit the potential of this technique. Indeed, we show the possible use of heatmap for packing or padding binary files analysis but there are many other

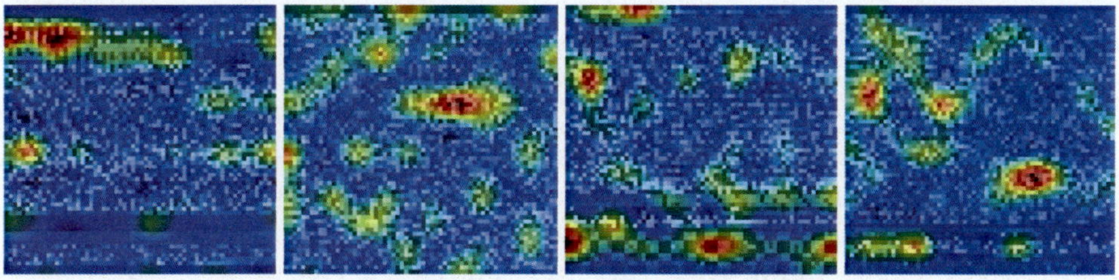

(a) Malware not packed (b) Malware packed (c) Benign not packed (d) Benign packed

Fig. 5. Attention map of some binary files

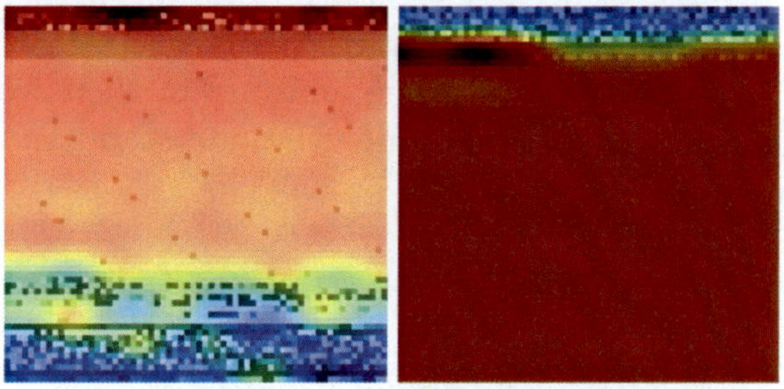

(a) Benign (b) Malware

Fig. 6. Example of binary files detected as malware due to padding

obfuscation techniques. Furthermore, we focus here on images of malware and benign, but an extension of this approach will be to extract code directly from the binary file based on the hot zone of the corresponding heatmap.

5 Conclusion and Results

Before concluding this paper, let us start by summarizing the results. In fact, CNN trained on RGB images based on HIT provides better results. However, the transformation time is too important to provide an efficient method for industrial deployment. Next, the DNN and LightGBM models demonstrate the expected effectiveness of the Ember dataset. Our models, which use grayscale images as input, are slightly less efficient than theirs. The results, however, are quite comparable to a training sample thirty times smaller. Finally, the two models train on grayscale images with information on whether the original binary file has undergone modifications or not pointed out the potential of this method for malware detection. They also provide more information on binary files than common detection models. CNN algorithm hybridized with RF, LGBM and SVM show an interesting detection potential. We will focus on this in future work to determine capacity or limit of this kind of models.

We have presented in this article different approaches to static malware detection. A recurring problem in many organizations is the computational time required for dynamic analysis of malware in a sandbox. However, with some modified malware, we know it is sometimes the only solution. We do not claim here to be able to replace this analysis, but we prefer proposing an overlay. Our algorithms enable us to quickly analyze a large amount of malware, determine which ones may or may not be malicious, while moderating the result if the binary is detected as modified or not. This can allow us to use dynamic analysis only on those binaries and, thus, save time and resources. In addition, analyzing the most important pixels and sections in image can also provide significant information for analysts. That can save them time in the detailed study of the suspicious binary by indicating where to look for it.

In future work, we will concentrate on attention mechanisms. The objective is to match the areas of importance, extracted with attention mechanisms, with the associated malicious code to help the analysts in their work. On the other hand, we want to use reinforcement learning to understand and prevent malware evasion mechanisms.

References

1. Sung, A.H., Xu, J., Chavez, P., Mukkamala, S.: Static analyzer of vicious executables. In: 20th Annual Computer Security Applications Conference, pp. 326–334. IEEE (2004)
2. Sathyanarayan, V.S., Kohli, P., Bruhadeshwar, B.: Signature generation and detection of malware families. In: Australasian Conference on Information Security and Privacy, pp. 336–349. Springer (2008)
3. Vasilescu, M., Gheorghe, L., Tapus, N.: Practical malware analysis based on sandboxing. In: Proceedings - RoEduNet IEEE International Conference, pp. 7–12 (2014)
4. Anderson, H.S., Roth, P.: Ember: an open dataset for training static PE malware machine learning models. arXiv preprint arXiv:1804.04637 (2018)
5. Raff, E., Barker, J., Sylvester, J., Brandon, R., Catanzaro, B., Nicholas, C.K.: Malware detection by eating a whole exe. arXiv preprint arXiv:1710.09435 (2017)
6. Fleshman, W., Raff, E., Sylvester, J., Forsyth, S., McLean, M.: Non-negative networks against adversarial attacks. arXiv preprint arXiv:1806.06108 (2018)
7. Nataraj, L., Karthikeyan, S., Jacob, G., Manjunath, B.S.: Malware images: visualization and automatic classification. In: Proceedings of the 8th International Symposium on Visualization for Cyber Security, pp. 1–7 (2011)
8. Vu, D.L., Nguyen, T.K., Nguyen, T.V., Nguyen, T.N., Massacci, F., Phung, P.H.: A convolutional transformation network for malware classification. In: 2019 6th NAFOSTED Conference on Information and Computer Science (NICS), pp. 234–239. IEEE (2019)
9. Deng, J., Dong, W., Socher, R., Li, L.J., Li, K., Fei-Fei, L.: Imagenet: a large-scale hierarchical image database. In: 2009 IEEE Conference on Computer Vision and Pattern Recognition, pp. 248–255. IEEE (2009)
10. Alom, M.Z., et al.: The history began from alexnet: a comprehensive survey on deep learning approaches. arXiv preprint arXiv:1803.01164 (2018)
11. Rezende, E., Ruppert, G., Carvalho, T., Ramos, F., De Geus, P.: Malicious software classification using transfer learning of resnet-50 deep neural network. In: 2017 16th IEEE International Conference on Machine Learning and Applications (ICMLA), pp. 1011–1014. IEEE (2017)

12. Yakura, H., Shinozaki, S., Nishimura, R., Oyama, Y., Sakuma, J.: Malware analysis of imaged binary samples by convolutional neural network with attention mechanism. In: Proceedings of the Eighth ACM Conference on Data and Application Security and Privacy, pp. 127–134 (2018)
13. Sharma, A., Sahay, S.K.: Evolution and detection of polymorphic and metamorphic malwares: a survey. Int. J. Comput. Appl. **90**(2), 7–11 (2014)
14. Zhang, Qinghua, Reeves, Douglas S.: MetaAware: identifying metamorphic malware. In: Proceedings - Annual Computer Security Applications Conference, ACSAC, pp. 411–420 2007 (2008)
15. Kreuk, F., Barak, A., Aviv-Reuven, S., Baruch, M., Pinkas, B., Keshet, J.: Adversarial examples on discrete sequences for beating whole-binary malware detection. arXiv preprint arXiv:1802.04528, pp. 490–510 (2018)
16. Aghakhani, H.: When malware is packin'heat; limits of machine learning classifiers based on static analysis features. In: Network and Distributed Systems Security (NDSS) Symposium 2020 (2020)
17. Christian Wojner. Bytehist. https://www.cert.at/en/downloads/software/software-bytehist
18. Chen, T., Guestrin, C.: Xgboost: a scalable tree boosting system. In: Proceedings of the 22nd ACM SIGKDD International Conference on Knowledge Discovery and Data Mining, pp. 785–794 (2016)
19. Ke, G.: Lightgbm: a highly efficient gradient boosting decision tree. In: Advances in Neural Information Processing Systems, vol. 30, pp. 3146–3154 (2017)
20. Xiao, Y., Xing, C., Zhang, T., Zhao, Z.: An intrusion detection model based on feature reduction and convolutional neural networks. IEEE Access **7**, 42210–42219 (2019)
21. Conti, G., et al.: A visual study of primitive binary fragment types. Black Hat USA, pp. 1–17 (2010)
22. Chattopadhay, A., Sarkar, A., Howlader, P., Balasubramanian, V.N.: Grad-CAM++: generalized gradient-based visual explanations for deep convolutional networks. In: 2018 IEEE Winter Conference on Applications of Computer Vision (WACV), pp. 839–847. IEEE (2018)

Check for updates

Byzantine Resilient Aggregation in Distributed Reinforcement Learning

Jiani Li(✉), Feiyang Cai, and Xenofon Koutsoukos

Institute for Software Integrated Systems, Vanderbilt University, Nashville, TN 37235, USA
{jiani.li,feiyang.cai,xenofon.koutsoukos}@vanderbilt.edu

Abstract. Recent distributed reinforcement learning techniques utilize networked agents to accelerate exploration and speed up learning. However, such techniques are not resilient in the presence of Byzantine agents which can disturb convergence. In this paper, we present a Byzantine resilient aggregation rule for distributed reinforcement learning with networked agents that incorporates the idea of optimizing the objective function in designing the aggregation rules. We evaluate our approach using multiple reinforcement learning environments for both value-based and policy-based methods with homogeneous and heterogeneous agents. The results show that cooperation using the proposed approach exhibits better learning performance than the non-cooperative case and is resilient in the presence of an arbitrary number of Byzantine agents.

1 Introduction

Due to the increasing volumes of data and the growth of machine learning (ML) applications, distributed learning and adaptation methods have been receiving greater attention. In such methods, multiple agents operate in a distributed and cooperative manner to achieve a common learning task. Typically, agents adapt their models using local data and interact with neighbors for model aggregation. Such cooperation has been demonstrated to help improve learning performance over the network [1]. Distributed reinforcement learning (RL), in particular, has been widely studied and applied in sensor networks, multi-robot networks, mobile phone networks, intelligent transportation systems, especially combined with deep neural networks [2–4].

Although cooperation in a distributed multi-agent network improves the learning performance, such methods are vulnerable to Byzantine attacks. It has been shown that a single Byzantine agent could disturb convergence of the entire network by sending malicious information to its neighbors [5,6]. To address this challenge, there is considerable recent research focusing on the resilient aggregation of distributed learning algorithms in the presence of Byzantine agents. Many resilient aggregation methods for distributed learning have been developed based on geometric properties of the model parameters such as coordinate-wise median, trimmed-mean, geometric median, Krum, and centerpoint, among many

© The Author(s), under exclusive license to Springer Nature Switzerland AG 2022
K. Matsui et al. (Eds.): DCAI 2021, LNNS 327, pp. 56–66, 2022.
https://doi.org/10.1007/978-3-030-86261-9_6

others [5,7–11]. One limitation of such approaches is that they are only resilient to a bounded number (usually less than half) of Byzantine neighbors.

Although research in Byzantine resilient aggregation for distributed ML is very broad, studies focusing on resilient distributed RL are limited. The recent studies in [12] and [13] use trimmed mean to achieve resilience for distributed actor-critic and Q-learning algorithms when a bounded number of agents are Byzantine. In this paper, we propose a Byzantine resilient aggregation rule for distributed RL. Compared to the existing methods that rely on the geometric properties to achieve resilience, the proposed method incorporates the idea of optimizing the objective function in designing the aggregation rule, and does not require a tailored upper bound of the Byzantine agents. In order to maximize the networked rewards, agents assign larger weights to neighbors incurring a larger reward, and stop cooperation with those incurring a smaller reward. Byzantine agents try to disturb the convergence of normal agents by sharing model parameters resulting in a small reward and thus are not taken into account by normal agents. The effectiveness of the proposed method is well validated by the evaluation results using multiple RL tasks for both value-based and policy-based distributed RL, such as distributed deep Q-learning and distributed Deep Deterministic Policy Gradient (DDPG). The evaluation results show that the proposed method exhibits better or similar learning performance (measured by the accumulated reward over the network) than no-cooperation in the presence of an arbitrary number of Byzantine agents.

2 Related Work

The technique of training multiple RL agents distributedly for a common objective has been extensively studied in recent years. Related work in distributed RL can be broadly grouped into two categories. In the first category, multiple agents operate in similar but independent Markov decision processes (MDPs) whose actions do not affect each other [14,15]. Such an approach is widely used in recent RL techniques for parallel exploration and computation to accelerate exploration and speed up learning, especially combined with deep neural networks. It is also naturally suited to be used in multi-agent networks where networked agents perform similar RL tasks in independent environments. The second category considers training RL algorithms with multiple agents in a single MDP [16]. This paper focuses on the first paradigm.

A major body of related work in training multiple agents in independent MDPs considers using a centralized parameter server for model updates and multiple workers to execute in multiple instances of the environment in parallel to collect state-action pairs and compute gradients of the model. Examples include the distributed deep Q-network [15] and A3C (Asynchronous Advantage Actor-Critic) [17]. Distributed RL in fully-decentralized networks has also been studied in the literature [14,18,19]. For example, [18] proposes a distributed implementation of Q-learning called QD-learning where every normal agent collaboratively updates tabular Q-values that being shared with their neighbors.

Further, [14] proposes a distributed RL method for policy evaluation with linear value function approximation. Moreover, [19] proposes a distributed actor-critic framework that aims to learn a policy that performs well on average for the whole set of tasks. Although research in Byzantine resilient aggregation for distributed learning algorithms is very broad, studies focusing on resilient distributed RL are limited. A recent work presented in [12] uses trimmed-mean to achieve resilience, where a centralized server exists in the network. In addition, a resilient version of QD-learning in a full-decentralized network has been proposed in [13], which is also based on the trimming approach.

3 Background

Markov decision processes (MDPs) are widely used for modeling RL problems, which can be described formally as a tuple $\mathcal{M} = \langle \mathcal{S}, \mathcal{A}, \mathcal{P}, R \rangle$, where $\mathcal{S}$ and $\mathcal{A}$ denote the (finite) state and action spaces, $\mathcal{P}(s'|s,a) : \mathcal{S} \times \mathcal{A} \times \mathcal{S} \to [0,1]$ is a state transition probability and $R(s,a,s') : \mathcal{S} \times \mathcal{A} \times \mathcal{S} \to \mathbb{R}$ is the reward function defined by $R(s,a,s') = \mathbb{E}[r_{t+1}|s_t = s, a_t = a, s_{t+1} = s']$ with r_{t+1} being the immediate reward received at time t. The probability of taking action a in state s is defined by the policy $\pi(a|s) : \mathcal{S} \times \mathcal{A} \to [0,1]$. Moreover, denote the state-value function $V_\pi(s) = \mathbb{E}_\pi \left[\sum_{t=0}^{\infty} \gamma^t r_{t+1} | s_0 = s, \pi\right]$, with $\gamma \in (0,1)$ as the discounted factor that determines how much future rewards are counted, and the action-value function $Q_\pi(s,a) = \sum_{s'} P_{ss'}^a (R(s,a,s') + \gamma V_\pi(s'))$, with $P_{ss'}^a = \mathcal{P}(s'|s,a)$. The objective is to learn an optimal policy π^* that maximizes the expected long-term reward given $a_t \sim \pi(\cdot|s_t)$ and $s_{t+1} \sim \mathcal{P}(\cdot|s_t, a_t)$:

$$\max_\pi \{J(\pi) = \mathbb{E}_s [V_\pi(s)]\}, \tag{1}$$

To ensure the existence of solution to (1), bounded rewards are assumed for any time-step as $|r_{t+1}| \leq r_{\max} < \infty, \forall t.$ for some scalar $r_{\max}$.

RL algorithms can be broadly categorized into value-based and policy-based. Below, we briefly introduce the main algorithms for the two types for solving (1).

Value-based methods aim to find a good estimate of the Q-function, and indirectly extract the optimal policy by selecting the greedy action in each state according to the estimates of the Q-values. One popular example is Q-learning [20], which uses the Bellman equation as an iterative update. Suppose $Q_\pi(s,a)$ can be parametrized by some parameter w as $Q(s,a;w) \approx Q_\pi(s,a)$. Then w can be updated by performing a gradient descent step on $\min_w \mathbb{E}\left[(y_t - Q(s_t, a_t; w_t))^2\right]$ where $y_t = \mathbb{E}\left[r_{t+1} + \gamma \max_{a'} Q(s_{t+1}, a'; w_{t-1}) | s_t, a_t\right]$ [21].

Policy-based methods directly search over the policy space to find the optimal policy instead of relying on the Q-function. One of the most popular policy-based RL algorithms is the Policy Gradient (PG) method [22]. In PG, the policy is parametrized by some parameter θ, and is updated by performing a gradient descent step on $\max_\theta J(\theta)$ with $\nabla J(\theta) = \mathbb{E}\left[\sum_{t=0}^{\infty} \Psi_t \nabla \log \pi_\theta(a_t|s_t)\right]$. Ψ_t can be expressed, for example, as the Q-function $Q_\pi(s_t, a_t)$ or the advantage function $A_\pi(s_t, a_t) = Q_\pi(s_t, a_t) - V_\pi(s_t)$, and can be estimated using Monte-Carlo evaluation [23].

4 Problem Formulation

Consider a network of $N + b$ agents operating in parallel based on similar but *independent* MDPs $\mathcal{M}_k = \langle \mathcal{S}_k, \mathcal{A}_k, \mathcal{P}_k, R_k \rangle$. The agents are connected by an *undirected graph* $\mathcal{G} = (\mathcal{V}, \mathcal{E})$ where $\mathcal{V}$ represents the agents and $\mathcal{E}$ represents interactions between agents. We assume that there are $b \geq 0$ Byzantine agents and $N \geq 1$ normal agents. *Normal* agents are those who strictly follow the prescribed algorithm in a network; and *Byzantine* agents are those who do not follow the algorithm and could send arbitrary different information to different neighbors usually with a malicious goal of disrupting the network's convergence. Note that Byzantine agents are *indistinguishable.* A bi-directional edge $(l, k) \in \mathcal{E}$ means that agents k and l can exchange information with each other. The *neighborhood* of k is the set $\mathcal{N}_k = \{l \in \mathcal{V} | (l, k) \in \mathcal{E}\} \cup \{k\}$. Agents share the same state and action spaces $\mathcal{S}_k$ and $\mathcal{A}_k$ but the transition probabilities $\mathcal{P}_k$ and the reward function R_k could be different among agents. Since agents are based on independent MDPs, their actions do not influence each other. Let J_k be the expected long-term reward associated with agent k. The goal is to cooperatively learn the optimal policies π_k for each normal agent k that maximize the global average reward:

$$\max_{\{\pi_k\}_{k=1}^N} \left\{ \frac{1}{N} \sum_{k=1}^{N} J_k(\pi_k) \right\}. \tag{2}$$

It is assumed that each normal agent k maintains its own parameter w_k (or θ_k), and uses $Q_k(s, a; w_k)$ (or $J_k(\theta_k)$) to be the local estimates of $Q_{\pi_k}(s, a)$ (or $J_k(\pi_k)$), when running value-based RL (or policy-based RL). Agents share their local estimates of such parameters with neighbors and aggregate the estimates from their neighbors to facilitate their learning. In this paper, we consider that the aggregation steps take place after each learning episode[1]. The algorithm used by each normal agent k with value-based or policy-based method for solving (2) is given in *Algorithm* 1[2].

Since Byzantine agents could disturb the convergence of normal agents through exchanging malicious messages, we are interested in finding a Byzantine resilient aggregation rule that solves (2) in the presence of Byzantine agents.

---

[1] An episode is a sequence of states from the start state to a terminal state.

[2] During learning, t increases while i remains the same; and during model aggregation, i increases while t changes from ∞ to 0. To simplify, hereafter, we omit the subscripts of t for cooperation and the superscripts of i for learning.

Algorithm 1: Distributed reinforcement learning with model aggregation

```
Input: Initialize w^0_{k,0} for value-based RL or θ^0_{k,0} for policy-based RL
for episode i = 0, M do
    Initialize state s^i_{k,0}, set t = 0 ;
    while s^i_{k,t} is not terminal do
        /* Exploration and learning                                   */
        Select a^i_{k,t} ~ π_k(·|s^i_{k,t})) ; Execute a^i_{k,t}, observe r^i_{k,t+1} and s^i_{k,t} ;
        Update w^i_{k,t} if using value-based method or θ_{k,t} if using policy-based
        method^3;
        t = t + 1 ;
    /* Model parameters aggregation                                   */
    Set w^i_{k,∞} = w^i_{k,t} or θ^i_{k,∞} = θ^i_{k,t}; Exchange w^i_{k,∞} or θ^i_{k,∞} with neighbors ;
    Assign weights c_{lk}(i) according to w^i_{l,∞} or θ^i_{l,∞} ;
    Aggregate estimates w^{i+1}_{k,0} = Σ_{l∈N_k} c_{lk}(i)w^i_{l,∞} or θ^{i+1}_{k,0} = Σ_{l∈N_k} c_{lk}(i)θ^i_{l,∞} ;
```

5 Resilient Aggregation in Distributed RL

In this section, we present resilient aggregation rules applicable to both value-based and policy-based distributed RL. The goal is to design non-negative weights $C_k(i) = [c_{1k}(i), \ldots, c_{(N+b)k}(i)] \in \mathbb{R}^{1\times(N+b)}$ for every normal agent k, with $\sum_{l\in\mathcal{N}_k} c_{lk}(i) = 1$, and $c_{lk}(i) = 0$ if $l \notin \mathcal{N}_k$, such that by using $C_k(i)$ in Algorithm 1, and in the presence of an arbitrary number of Byzantine neighbors, (2) is solved. We incorporate the idea of optimizing the objective function in designing the aggregation weights. Besides, a softmax layer is applied to the weights in order to make the weights to be non-negative. The weights are assigned as follows: if $l \notin \mathcal{N}_k^{\geq}$, then $c_{lk}(i) = 0$, otherwise,

$$c_{lk}(i) = \frac{e^{\hat{J}_k(\pi_l^i)}}{\sum_{p\in\mathcal{N}_k^{\geq}} e^{\hat{J}_k(\pi_p^i)}}, \tag{3}$$

where $l \in \mathcal{N}_k^{\geq}$ if $l \in \mathcal{N}_k$ and $\hat{J}_k(\pi_l^i) \geq \hat{J}_k(\pi_k^i)$; $\hat{J}_k(\cdot)$ is an approximation of $J_k(\cdot)$ with $\mathbb{E}_s[\hat{J}_k(\cdot)] = J_k(\cdot)$. For example, $\hat{J}_k(\pi)$ can be computed by the simulated long-term reward of one-shot Monte-Carlo policy evaluation using policy π on the MDP of k. Note that π_l^i can be either extracted from w_l^i when using value-based RL or can be parametrized by θ_l^i when using policy-based RL. Obviously, $\sum_{l\in\mathcal{N}_k} c_{lk}(i) = 1$ holds using weights (3). The intuition behind (3) is the following. One agent k can evaluate the policy of a neighbor l on its own MDP, and a larger long-term reward resulted by a neighbor's policy on k's MDP implies a better approximation of the policy and agent k should assign larger weights to

[3] Methods for updating these parameters are discussed in Sect. 3.

such policies. In addition, it is reasonable to cooperate with neighbors incurring better approximations but unnecessary to cooperate with those resulting in worse approximations than oneself. As a result, we come up with the aggregation rule given in (3).

Discussion of the Convergence. Convergence for RL algorithms is hard to be guaranteed, especially when combined with neural networks. In the convergence analysis of RL algorithms, it is often assumed that the value or policy functions can be parametrized by a class of linear functions. If this is the case, the aggregation using (3) results in $\hat{J}_k(\pi_k^{i+1}) \geq \hat{J}_k(\pi_k^i)$, given that the agent cooperates only with the neighbors having a larger approximate reward $\hat{J}_k(\pi_l^i) \geq \hat{J}_k(\pi_k^i)$, and $\hat{J}_k(\pi_k^{i+1})$ is a linear combination of $\hat{J}_k(\pi_l^i)$. This means that the aggregation using (3) always results in a larger approximate long-term reward, i.e., a better approximation of the policy. Further, the presence of Byzantine agents does not affect this improvement. As a result, if agents converge to the optimal policy without cooperation, they also converge in the cooperative case in the presence of byzantine agents. If agents do not converge to the optimal policy without cooperation, cooperation using (3) could help improve their learned policy.

6 Evaluation

In this section, we evaluate the proposed resilient aggregation method for both value-based and policy-based RL algorithms. We also compare the approach with the average- and median-based aggregation rules as well as the non-cooperative case. Note that median is a special case of trimmed-mean when half of the smallest and largest values are trimmed. In all the examples, our approach exhibits better or similar learning performance than the non-cooperative case measured by the averaged long-term rewards over the network, in the presence of an arbitrary number of Byzantine neighbors. When all the neighbors are Byzantine, the approach is reduced to a non-cooperative algorithm. Whereas in the same scenarios, average and median-based methods may exist worse learning performance than the non-cooperative case, showing the vulnerabilities of such aggregation methods in Byzantine systems.

6.1 Simulation Setup

Tasks. The tasks we consider are the classic control problems Cartpole and Pendulum, and the Atari game Pong based on the OpenAI Gym [24]. Specifically, we evaluate both the value-based RL algorithm deep Q-networks (DQN) [21] for Cartpole and Pong, and policy-based algorithm DDPG [25] for Pendulum.

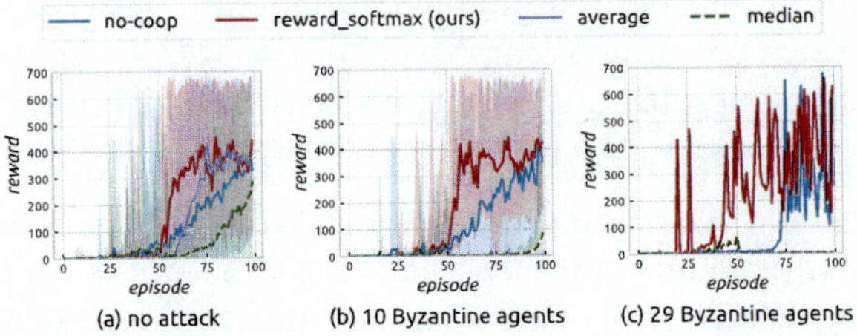

Fig. 1. 30 homogeneous agents running DQN for Cartpole.

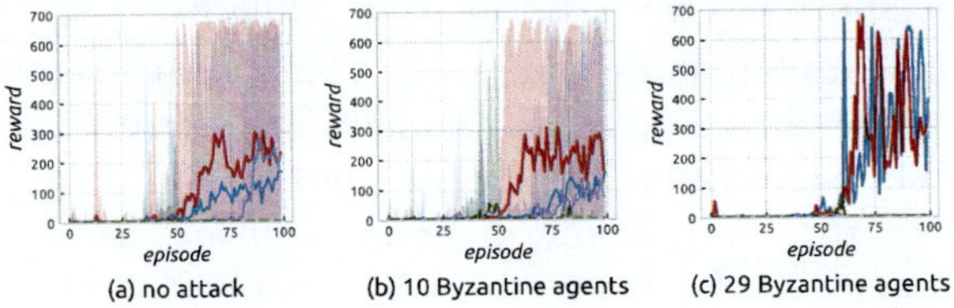

Fig. 2. 30 heterogeneous agents running DQN for Cartpole.

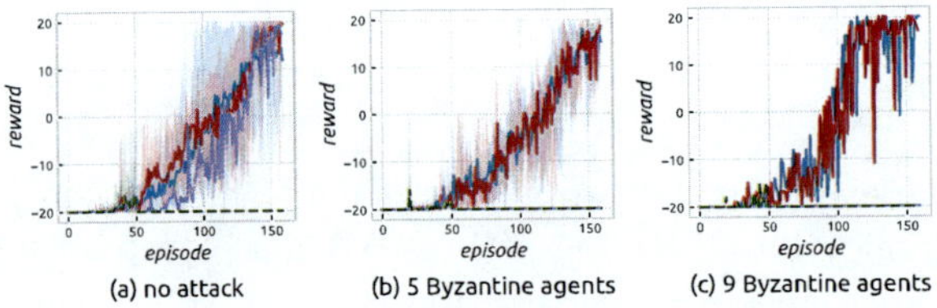

Fig. 3. 10 homogeneous agents running DQN for Pong.

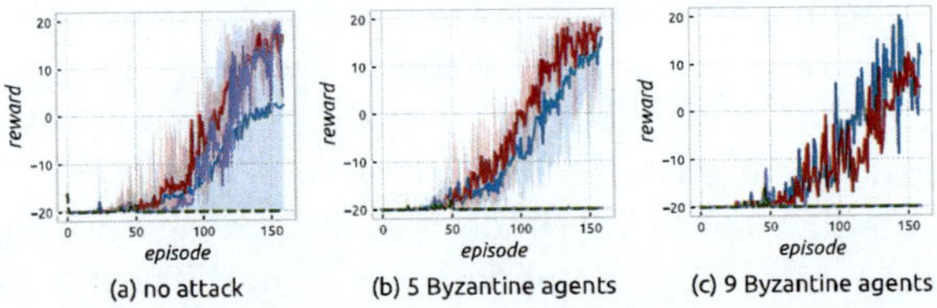

Fig. 4. 10 heterogeneous agents running DQN for Pong.

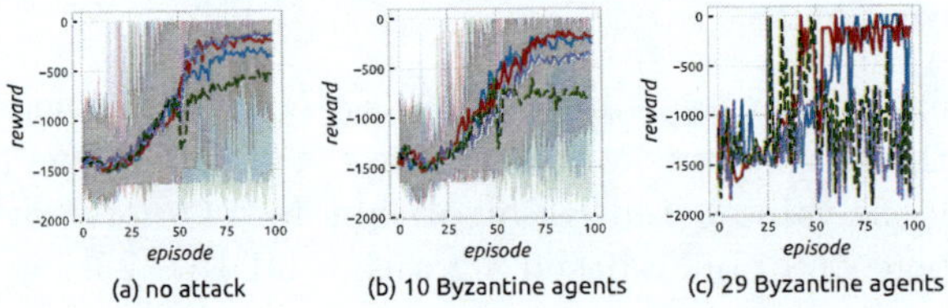

Fig. 5. 30 homogeneous agents running DDPG for Pendulum.

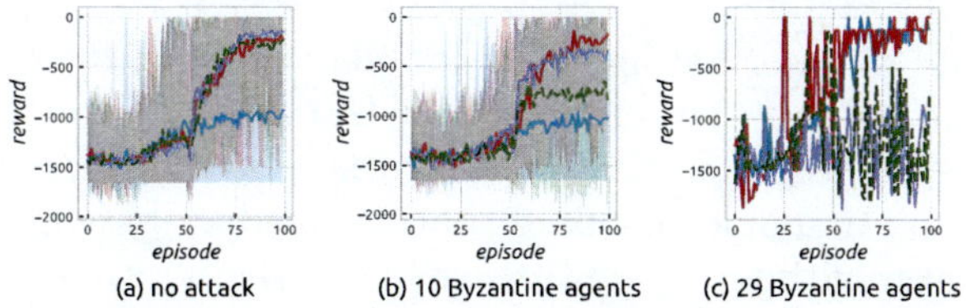

Fig. 6. 30 heterogeneous agents running DDPG for Pendulum.

Network. For the Cartpole and Pendulum tasks, we consider a network of 30 agents; and for Pong, we consider a network of 10 agents. The connectivity of the agents is determined by their geographical location, which is randomly drawn from a $[0, 3] \times [0, 3]$ plane. The average degree $\frac{1}{N}\sum_{k=1}^{N}|\mathcal{N}_k|$ of the connectivity graph with 30 agents is approximately 8.2. The network with 10 agents is modeled by a complete graph.

Attacks. Byzantine agents are designed to send random values (for each dimension) from the interval [0, 1] to all of its neighbors.

Hyper-parameters. In all the tasks, we use ADAM optimizer [26], $\gamma = 0.99$, and ε-greedy exploration strategy with ε annealed from 1 to 0.01 over the first 1e5 learning steps [15] for DQN. The aggregation between the agents starts after the agents run 50 episodes' exploration and learning independently. For Cartpole, the batch-size is 32 and the neural network model for each agent has one hidden layer of 50 neurons. For Pong, the batch-size is 32 and we use CNN with 3 convolutional layers of output shape $[32, 20, 20]$, $[64, 9, 9]$ and $[64, 7, 7]$, as well as a fully-connected layers of 512 neurons and an output layer. For Pendulum, the batch-size is 128 and the actor network has 3 hidden layers with 256/128/64 neurons; The critic network has two layers with 256/128 neurons for the state input and one layer with 128 neurons for the action input, which then concatenates the state/action output of 256 neurons followed by a fully-connected layer of 128 neurons and an output layer. The activation functions are all ReLU. In a connected network, every agent shares the same model architecture (and the same state/action spaces).

6.2 Simulation Results

We consider both homogeneous networks where agents use the same learning rate and environment as well as heterogeneous networks where the agents have different learning rates and/or random noisy data being added to each state.

Homogeneous Agents. We set the same learning rate and other parameters for each normal agent. The learning rate is 0.01 for Cartpole, 0.001 for Pendulum, and 0.0001 for Pong. Figure 1, 3 and 5 show the mean and range of the evaluated accumulated reward for every normal agent in the network, in the case of no attack, with 5/10 Byzantine agents, and with 9/29 Byzantine agents. We find that the proposed aggregation method is resilient in all scenarios (even when there is only one normal agent in the network and the other agents are all Byzantine) and exhibits better or similar learning performance as the non-cooperative case. The average and median rules fail to converge in the presence of Byzantine agents. It should be noted that the median-based aggregation may fail to converge even without Byzantine agents in the network.

Heterogeneous Agents. In the heterogeneous networks, we consider a random learning rate sampled from $(0, 0.1]$ for Cartpole; from $[0.0001, 0.0002]$ for Pong; and from $(0, 0.01]$ for Pendulum. We also consider random noise sampled from $(0, 0.02]$ being added to each element of the state for Cartpole and Pendulum. Figure 2, 4 and 6 show the results. The proposed method outperforms no-cooperation, average, and median as measured by the average accumulated rewards over the network, with and without Byzantine agents. Comparing to the results of the homogeneous setting, we find that in heterogeneous networks, agents that are not able to reach a good policy by themselves could greatly improve their learning performance by cooperating with neighbors. In general, the averaged accumulated reward over the network is greatly improved by model aggregation using the proposed weights compared to the non-cooperative case.

7 Conclusion

In this paper, we present a Byzantine resilient aggregation rule for distributed reinforcement learning with networked agents. In order to maximize the networked rewards, agents assign larger weights to neighbors incurring a larger reward and reduce cooperation with those incurring a smaller reward. Byzantine agents try to disturb the convergence of normal agents and share a model resulting in small rewards, and thus, they are not included in the cooperation with normal agents. Compared to the previous methods that relies on a tailored upper bound of the Byzantine agents in achieving resilience, the proposed method does not require such a bound and can be resilient even when all the neighbors of a normal agent are Byzantine. We evaluate our approach using multiple RL tasks, for both value- and policy-based methods, with homogeneous and heterogeneous agents. The simulation results validate the effectiveness of our approach, showing that cooperation using the proposed approach improves the learning performance over the network, in the presence of an arbitrary number of Byzantine agents.

References

1. Sayed, A.H., Tu, S.Y., Chen, J., Zhao, X., Towfic, Z.J.: Diffusion strategies for adaptation and learning over networks: an examination of distributed strategies and network behavior. IEEE Signal Process. Mag. **30**(3), 155–171 (2013)
2. Zhang, K., Yang, Z., Liu, H., Zhan g, T., Basar, T.: Fully decentralized multi-agent reinforcement learning with networked agents. In: ICML 2018, Stockholmsmässan, Stockholm, Sweden, 10–15 July 2018, pp. 5867–5876 (2018)

3. Mnih, V., et al.: Asynchronous methods for deep reinforcement learning. In: JMLR Workshop and Conference Proceedings of, ICML 2016, New York City, NY, USA, 19-24 June 2016, vol. 48, pp. 1928–1937. JMLR.org (2016)
4. Espeholt, L., et al.: IMPALA: scalable distributed Deep-RL with importance weighted actor-learner architectures. In: ICML 2018, Stockholm, Sweden, 10-15 July 2018
5. Blanchard, P., El Mhamdi, E.M., Guerraoui, R., Stainer, J.: Machine learning with adversaries: byzantine tolerant gradient descent. In: Annual Conference on Neural Information Processing Systems, pp. 118–128 (2017)
6. Li, J., Abbas, W., Koutsoukos, X.: Resilient distributed diffusion in networks with adversaries. IEEE Trans. Signal Inf. Process. over Netw. **6**, 1–17 (2020)
7. Yin, D., Chen, Y., Kannan, R., Bartlett, P.: Byzantine-robust distributed learning: towards optimal statistical rates. In: ICML 2018, Stockholmsmässan, Stockholm, Sweden, 10-15 July 2018, pp. 5636–5645 (2018)
8. Yang, Z., Bajwa, W.U.: ByRDiE: byzantine-resilient distributed coordinate descent for decentralized learning. IEEE Trans. Signal Info. Process. Over Netw. **5**(4), 611–627 (2019)
9. Chen, Y., Su, L., Xu, J.: Distributed statistical machine learning in adversarial settings: byzantine gradient descent. In: Proceedings of the ACM on Measurement and Analysis of Computing Systems, vol. 1, no. 2, pp. 44:1–44:25, December 2017
10. Li, J., Abbas, W., Shabbir, M., Koutsoukos, X.: Resilient distributed diffusion for multi-robot systems using centerpoint. In: Proceedings of Robotics: Science and Systems, Corvalis, Oregon, USA, July 2020
11. Li, J., Abbas, W., Koutsoukos, X.: Byzantine resilient distributed multi-task learning. In: Advances in Neural Information Processing Systems 33: Annual Conference on Neural Information Processing Systems, 6-12 December 2020 (2020)
12. Lin, Y., Gade, S., Sandhu, R., Liu, J.: Toward resilient multi-agent actor-critic algorithms for distributed reinforcement learning. In: 2020 American Control Conference, ACC 2020, Denver, CO, USA, 1-3 July 2020, pp. 3953–3958. IEEE (2020)
13. Xie, Y., Mou, S., Sundaram, S.: Towards resilience for multi-agent QD-learning. CoRR, abs/2104.03153 (2021)
14. Macua, S.V., et.al.: Distributed policy evaluation under multiple behavior strategies. IEEE Trans. Automat. Contr. **60**(5), 1260–1274 (2015)
15. Nair, A., et al.: Massively parallel methods for deep reinforcement learning. CoRR, abs/1507.04296 (2015)
16. Zhang, K., Yang, Z., Basar, T.: Multi-agent reinforcement learning: a selective overview of theories and algorithms. CoRR, abs/1911.10635 (2019)
17. Balcan, M.F., Weinberger, K.Q. (eds.) Proceedings of the 33nd International Conference on Machine Learning, ICML 2016, New York City, NY, USA, 19-24 June 2016, vol. 48 of JMLR Workshop and Conference Proceedings. JMLR.org (2016)
18. Kar, S., Moura, J.M., Poor, H.V.: QD-learning: a collaborative distributed strategy for multi-agent reinforcement learning through Consensus + Innovations. IEEE Trans. Signal Process. **61**(7), 1848–1862 (2013)
19. Macua, S.V., Tukiainen, A., Hernández, D.G.O., Baldazo, D., de Cote, E.M., Zazo, S.: Diff-dac: distributed actor-critic for multitask deep reinforcement learning. CoRR, abs/1710.10363 (2017)
20. Watkins, C.J., Dayan, P.: Q-learning. In: Machine Learning, pp. 279–292 (1992)
21. Mnih, V., et al.: Playing atari with deep reinforcement learning. CoRR, abs/1312.5602 (2013)

22. Sutton, R.S., McAllester, D.A., Singh, S.P., Mansour, Y.: Policy gradient methods for reinforcement learning with function approximation. In: Advances in Neural Information Processing Systems 12, Denver, Colorado, USA, pp. 1057–1063 (1999)
23. Schulman, J., Moritz, P., Levine, S., Jordan, M., Abbeel, P.: High-dimensional continuous control using generalized advantage estimation. In: ICLR 2016, San Juan, Puerto Rico, 2-4 May 2016, Conference Track Proceedings (2016)
24. Brockman, G.: OpenAI Gym. arXiv preprint arXiv:1606.01540 (2016)
25. Lillicrap, T.P., et al.: Continuous control with deep reinforcement learning. In: ICLR (2016)
26. Kingma, D.P., Ba, J.: Adam: a method for stochastic optimization. In: ICLR 2015, San Diego, CA, USA, 7-9 May 2015

Utilising Data from Multiple Production Lines for Predictive Deep Learning Models

Niclas Ståhl(✉), Gunnar Mathiason, and Juhee Bae

University of Skövde, Kanikegränd 3a, 541 34 Skövde, Sweden
niclas.stahl@his.se

Abstract. A Basic Oxygen Furnace (BOF) for steel making is a complex industrial process that is difficult to monitor due to the harsh environment, so the collected production data is very limited given the process complexity. Also, such production data has a low degree of variability. An accurate machine learning (ML) model for predicting production outcome requires both large and varied data, so utilising data from multiple BOFs will allow for more capable ML models, since both the amount and variability of data increases. Data collection setups for different BOFs are different, such that data sets are not compatible to directly join for ML training. Our approach is to let a neural network benefit from these collection differences in a joint training model.

We present a neural network-based approach that simultaneously and jointly co-trains on several data sets. Our novelty is that the first network layer finds an internal representation of each individual BOF, while the other layers use this representation to concurrently learn a common BOF model. Our evaluation shows that the prediction accuracy of the common model increases compared to separate models trained on individual furnaces' data sets. It is clear that multiple data sets can be utilised this way to increase model accuracy for better production prediction performance. For the industry, this means that the amount of available data for model training increases and thereby more capable ML models can be trained when having access to multiple data sets describing the same or similar manufacturing processes.

Keywords: Deep learning · Data fusion · Joint training · Steel making

1 Introduction

Basic oxygen steelmaking in a Linz-Donawitz-converter (LD-converter) furnace is a complex chemical and physical industrial process that reduces a mix of pig iron and recycled scrap into low-carbon steel in a high-temperature furnace. This process has many complex dependencies between the process parameters and the final result, but only a few measured parameters are available from sensors during the process due to the harsh environment.

© The Author(s), under exclusive license to Springer Nature Switzerland AG 2022
K. Matsui et al. (Eds.): DCAI 2021, LNNS 327, pp. 67–76, 2022.
https://doi.org/10.1007/978-3-030-86261-9_7

A large share of the collected process data represents well-controlled process executions that are all very similar in their data distributions. However, a machine learning approach aimed to capture the dependencies of a complex process requires training data that contains distributions with a full spectrum of various process executions. Limited distribution and resolution, caused by the lack of variation and the lack of in-process sensor data, limits the understanding of the physical process in the trained model. For a model to fully capture the furnace process complexity, there is a need to make use of as much of the collected data as possible. Using a combination of data from multiple furnaces allows for both more and richer training data. With better data and more complex machine learning, more unknown dependencies of the actual process can be captured in a model. Moreover, such a model could also find generic dependencies that exist in the typical furnace process.

We introduce an approach for learning from process data that are collected from multiple furnaces, where the trained model can benefit from complimentary data collection. With this approach, the information about each furnace can be integrated into a joint machine learning model. For evaluation of the presented machine learning model, we use data from three furnaces at two different sites, with different data collection setups at each site and furnace.

The aim of the presented model is to achieve higher utilisation of the data available, and to utilise the information that is available about the physical process in the different plants. The model captures the specific execution patterns from all three furnaces into one generic model, capable of predicting the process outcome for all furnaces. Further, the variation of three data collection setups enriches the training data set, since a wider range of different process executions are considered. Furthermore, a joint training model is trained on more data, data with a better and enriched representative distribution.

The model for joint training, using data from the collected data sets, an artificial neural network that is structured to support co-training. In the network, we separate the generic training layers of the model from a first furnace-specific training layer, such that all but the first layer in the model are generic and shared for all three data sets. The idea behind this structure is that the first layer, which is specific to each furnace, transforms the input data into a common abstraction of the BOF process that then can be used by the following generic layers.

When evaluating the joint model on our collected data, the prediction accuracy increases compared to having three separate models trained for an individual furnace's data set. Thus, we see that the joint model as we propose it better captures the complex behaviour of the actual steel manufacturing process in the furnace, in spite of the sparseness of the data available for monitoring such a process.

2 Background

The joint training aspect of our approach is inspired by *transfer learning* (Torrey and Shavlik 2010), which is a frequently used approach when only a limited dataset is available for a specific problem of too high complexity given the available data. Transfer learning has been shown to work well in many domains with limited data. For instance, it is used in language translation where a limited language-specific training set is available (Luo et al. 2019). In transfer learning, a more generic model is first trained on a large dataset in a domain that is similar to the targeted problem. For language training, the generic model can be pre-trained on another language, or a set of languages, where more labelled data is available. The trained model is re-used (transferred) to the more specific problem, such that model configuration is fine-tuned using the smaller but more problem-specific dataset. Our case is different in that there is no large common dataset in the domain that can be used for pre-training. Our approach is similar in that we expect latent and common information in the data that represents most of the knowledge needed for a model. Instead of pre-training, we learn the latent and common knowledge in the available data by learning a common abstraction for several smaller data sets, such that the smaller data sets in combination replace a large dataset.

There is a body of work on how to predict the process targets in a BOF process (Bae et al. 2020; Viana Junior et al. 2018; Li et al. 2016; Laha et al. 2015; Bing-yao et al. 2011, Wang et al. 2010; Han and Jiang 2010; Cox et al. 2002). The various approaches and cases configure their methods in various ways, such as the size and fidelity of the dataset, the chosen ML algorithm, variability and size of the datasets, the target error range and the number of used features. These previous works show that the choices of data and algorithms determine prediction accuracy, and we argue that only a full distribution data set will be representative and should be used for training a useful prediction model.

Prediction accuracy is to a large extent influenced by how narrow the parameter choice is made. A common pattern is that more complex machine learning algorithms cannot utilise small data sets with too little training example distribution. In such cases, the model only focuses on remembering simplistic patterns and the data is not fully utilised. With richer data that has a distribution more truly to the actual process distribution, such as when the process is represented with a larger and better distribution, more advanced machine learning algorithms can benefit and capture the actual process complexity. This results in models that are more successful with validation examples from actual production.

3 Method

This section first describes the data that is used for the experiments and secondly the model that is used. The experiment setup and all parameters are described at the end of the section.

3.1 Data

Data are collected from an actual production in an industrial LD-converter, run by the SSAB group. The data used in this paper are collected from three different BOF production lines, which are distributed at two different production sites, both located in Sweden. Hence, there are three different data sets, where the data sets have some overlapping features but also some features that are unique to each data set. Some are collected in an identical way, while others have a different representation. Further, some features are not collected at all for some furnace, while some furnace has a unique and detailed data collection for certain parameters lacking for the others. There are 55, 88 and 89 features that are measured and stored in the three data sets, respectively. These features include parameters before the oxygen blow and during the oxygen blow. Some parameters represent the statistical descriptors (e.g., difference) of the time-series dataset collected throughout the oxygen blow. Others include thermodynamic engineered parameters using the originally collected parameters applied to thermodynamic principles.

There are three targets: temperature, carbon, and phosphorus. Similar to the target for the physical process, the aim of the model is to predict the end temperature (T) as well as the percentage of carbon (C) and phosphorus (P) in the final product. Thus, all three data sets contain measurements of the resulting values for these targets. With the physical process, the aim is that in each process execution (a *heat*) the steel temperature should be within a certain temperature range, where also carbon and phosphorus levels are lower than specified thresholds. To be able to improve the physical process control, operators benefit from having a good prediction of the expected execution time, given all the data that is known for each specific heat. These three data sets contain 10,406, 22,908 and 22,758 samples respectively, which gives a total of 56,072 heats. From the heats collected, a number of heats were removed that had missing and corrupted values. The initial heats for tuning and the heats outside of the blow time threshold were regarded as outliers and were removed which the criteria were decided by the domain experts. All heats come from continuous production during the investigated time. When the data are used to train the model, the data are first normalised so that every feature has a mean of zero and where each sample is measured in the standard deviations from the original mean. This normalisation process is conducted independently in each of the three datasets.

3.2 Model

The model, fully depicted in Fig. 1, is a model based on an artificial neural network. All layers of the neural network are shared except the initial hidden layer, even the last output layer, which is used for predictions, is shared. The first layer of the neural network is specific for each dataset and the idea is that these layers perform a transformation of the data into a hidden abstract representation of the BOF-process that is common for all three data sets and should contain the state of the process regardless of which furnace the data come from. These hidden representations are propagated through a joint neural network and predictions are made by the last layer. The neurons in the output layer do not use any nonlinear activation function, but only propagate the linear transformation of the input signals. All other neurons in the network use a leaky rectified linear unit (ReLU) as the activation function (Maas et al. 2013).

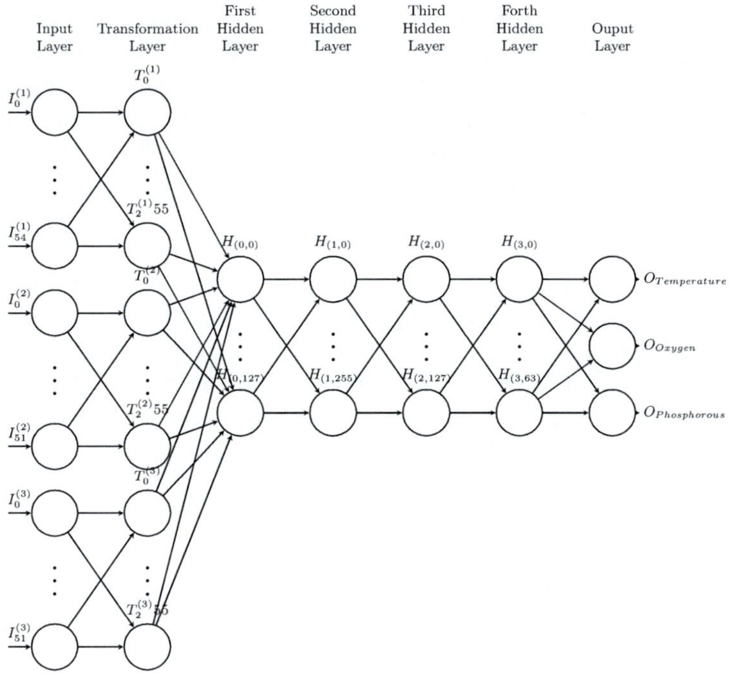

Fig. 1. The full model, for analysing data from multiple production lines. The first layer consists of input from three different sources. These are denoted $I_j^{(i)}$ where i specify the input source and j the number of the feature from that source. The different inputs are first propagated through a production line-specific layer of neurons. This is called the transformation layer and each neuron in such a layer is denoted $T_j^{(i)}$ where i denotes the production line and j is the numbering of the neuron in that layer. The output of the transformation layers is then propagated through four shared hidden layers. The neurons in these layers are denoted $H_{(i,j)}$ where the j:th neuron in the i:th hidden layer. These layers are followed by the final output layer. This layer contains three neurons, outputting a prediction for each of the targeted features for the given input.

3.3 Experiment

To evaluate the improvement of having a jointly trained prediction model, we train four types of models: a neural network-based joint model; individual single-furnace models with a similar network design; a random forest-based model; and a support vector machine model. Here the three later of the models are all applied to the three data sets independently and cannot utilise any connection or correlations between the data sets.

The individual model is a neural network with the same network design as the proposed joint model in Fig. 1, except that there are no shared layers and, hence, the transformation layer would just be a regular hidden layer. All other parameters are the same as in the joint model. Both these neural network-based models are optimised using the ADAM optimisation algorithm (Kingma and Ba 2014) with a learning rate of 0.001 and a batch size of 512 samples. A 20% dropout rate (Srivastava et al. 2014) is introduced to all connections in the network during the training phase to prevent the network from overfitting.

They are trained for 100 epochs, where the final model that is selected is the model that performs best on a separate validation set.

We choose to compare the neural network-based results with a random forest model containing 100 trees as well as a support vector machine-based model. These types of models have been shown to perform equally well as neural networks for BOF-target predictions based on our data sets (Bae et al. 2020).

A 10-fold cross validation is used to evaluate the results of the three different methods. For the two neural networks methods, eight of the ten folds are used for training the model, one is used for model validation and the last one is used as the test set, for which the result is reported. For the random forest and the support vector machine, which does not utilise a validation set, nine of the folds are used for training and the final set is used for testing.

4 Result

To quantify the predictive power of the presented model, as well as the ones used for comparison, the R^2 score is measured for the predictions over all test sets in the 10-fold cross validation. These results are presented in Table 1. The results in this table show that the joint model performs better than the other two models in all cases and for all three data sets. These results are further dissected by plotting the normalised real values from each dataset over the predictions that are made by the joint model. This is shown in Fig. 2 and in this figure the difference in the R^2 scores between 0.59 and 0.70 in the temperature case and the R^2 scores of 0.11–0.25 in the carbon prediction case.

Table 1. The R^2 score for the three different datasets, from different production lines, for all three target features that are predicted. The highest R^2 score for each dataset and feature is presented in bold font.

	Joint neural network	Neural network	SVM	Random forest
Temperature	**0.69**	0.59	0.68	0.60
	0.67	0.61	0.52	0.62
	0.67	0.63	0.50	0.64
Carbon	**0.26**	0.20	0.24	0.21
	0.49	0.39	0.17	0.35
	0.48	0.38	0.18	0.37
Phosphorus	**0.42**	0.27	0.40	0.32
	0.64	0.58	0.44	0.60
	0.63	0.55	0.43	0.58

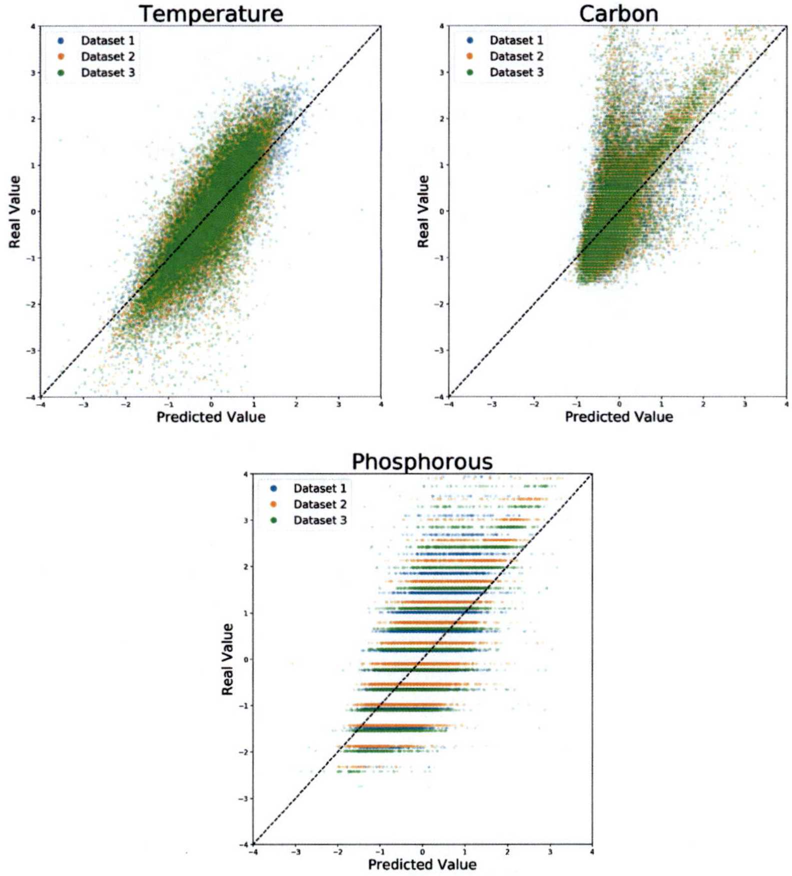

Fig. 2. Predicted vs. actual values on three targets temperature, carbon, and phosphorus

5 Discussion and Conclusion

In our previous work (Bae et al. 2020), we explored the suitability of common machine learning algorithms to capture the complexity of a Basic Oxygen Furnace, using a full data set from years of production, for predicting a critical combined production target. Each of the top-performing algorithms found reached similar or better results than existing work found in literature, even for those that used an advantageous selection of data. We interpreted the similarity of our prediction results from different approaches that the data we used could not give more training information, and any training model would be limited for further prediction improvements on this data. Data is typically limited by the inherent uncertainty of the data collected that stems from both stochastic process components, sensor setup uncertainty, and sensor limitations.

This paper explores the fact that multiple data sets coming from different furnaces, here in the same company, should compensate for some uncertainties caused by data collection limitations and the sparseness of information. The main contribution of this paper is the approach to explore that for even better process predictions. We expand one of our previous top-performing approaches, the neural network-based learning algorithm, into a deeper network that can jointly train on multiple data sets to learn a furnace-unique representation that becomes significant to improving the prediction of the combined multi-class process target. The improved results show that there is a clear benefit to let a learning system find a latent representation of individual complementing data sets, and concurrently also train on several such representations in a joint training model. Because of the learning of a hidden latent representation, which is common to the process, the joint model can be trained well even when the set of training features are different for each data set.

Our results show that the joint model with such a network design performs better (Table 1). It achieves a higher R^2 score than all three alternative approaches, for all three data sets, and for all prediction targets. The least improvement can be seen for the R^2 scores for the carbon target prediction. This is a consequence of the low variability and granularity in the sensor measurements. In the current process control system, carbon is well-regulated since it is the most critical target, and a consequence is that there is low variability in the training examples since many of the furnace heats are successful. Achieving these R^2 scores in a joint training approach, roughly 64% of all predictions can be used for improving the current success rate in production. Predictions need to be accurate enough, so the prediction error needs to be within: 15 °C for the temperature prediction, 2% in the carbon content prediction and 0.3% in the phosphorus content prediction.

A bi-modality in the predictions can be observed when further analysing the predicted values for carbon in relation to the observed values, which is depicted in Fig. 2. This bi-modality arises due to the fact that not all events, which have an effect on the outcome during production, are recorded and provided to the model. Hence, in some cases the process is affected by one or several events that are not known to the model. When the model gets such a sample without all crucial information, it cannot predict it well and the prediction will be as if the event did not occur during the production. Such a sample would correspond to a point in the leftmost cluster. On the other hand, we can see that all information that is needed to predict samples well is provided for the samples in the rightmost cluster. The same effect can also be observed in the phosphorous prediction, while not as distinct.

Our results show that the joint training approach that we suggest can find a representation of the properties of the similarity of the physical manufacturing process. Even for differently collected data sets for this process, the combination can be utilised in a joint training neural network. By letting the first layer learning a transformation of specifics of each individual data set into a shared representation, the rest of the model can be trained on abstractions of the original data sets. The representation of each data set is found by training the model and does not need to be designed by a human expert based on what is known about differences in the data collection. For the industry, this means that the amount of available data can be increased and better models can be trained for data describing the same manufacturing process coming from different sites.

Acknowledgements. We would like to thank Niklas Kojola, Carl Ellström, Patrik Wikström, and Lennart Gustavsson at SSAB for the close collaboration in the Swedish Metal project under which this research is funded. The project is funded by the Knowledge Foundation in Sweden, under the grant 20170297.

References

Torrey, L., Shavlik, J.: Transfer learning. In Handbook of Research on Machine Learning Applications and Trends: Algorithms, Methods, and Techniques, pp. 242–264. IGI global (2010)

Luo, G., Yang, Y., Yuan, Y., Chen, Z., Ainiwaer, A.: Hierarchical transfer learning architecture for low-resource neural machine translation. IEEE Access **7**, 154157–154166 (2019). ISSN 2169-3536. Conference Name: IEEE Access

Bae, J., Li, Y., Ståhl, N., Mathiason, G., Kojola, N.: Using machine learning for robust target prediction in a Basic Oxygen Furnace system. Metall. Mater. Trans. B **51**, 1632–1645 (2020)

Viana Junior, M.A., Silva, C.A., Silva, I.A.: Hybrid model associating thermodynamic calculations and artificial neural network in order to predict molten steel temperature evolution from blowing end of a BOF for secondary metallurgy. REM Int. Eng. J. **71**(4), 587–592 (2018). ISSN 2448-167X

Li, W., Wang, X., Wang, X., Wang, H.: Endpoint prediction of BOF steelmaking based on BP neural network combined with improved PSO. Chem. Eng. Trans. **51**, 475–480 (2016)

Laha, D., Ren, Y., Suganthan, P.N.: Modeling of steelmaking process with effective machine learning techniques. Expert Syst. Appl. **42**(10), 4687–4696 (2015). ISSN 0957-4174

Bing-yao, C., Hui, Z., You-jun, Y.: Research on the BOF steelmaking endpoint temperature prediction. In: 2011 International Conference on Mechatronic Science, Electric Engineering and Computer (MEC), pp. 2278–2281, August 2011

Wang, X., Han, M., Wang, J.: Applying input variables selection technique on input weighted support vector machine modeling for BOF endpoint prediction. Eng. Appl. Artif. Intell. **23**(6), 1012–1018 (2010). ISSN 0952-1976

Han, M., Jiang, L.: Endpoint prediction model of basic oxygen furnace steelmaking based on PSO-ICA and RBF neural network. In: 2010 International Conference on Intelligent Control and Information Processing, pp. 388–393, August 2010

Cox, I.J., Lewis, R.W., Ransing, R.S., Laszczewski, H., Berni, G.: Application of neural computing in basic oxygen steelmaking. J. Mater. Process. Technol. **120**(1), 310–315 (2002). ISSN 0924-0136

Maas, A.L., Hannun, A.Y., Ng, A.Y.: Rectifier nonlinearities improve neural network acoustic models. In: Proceedings of ICML, vol. 30, p. 3 (2013)

Kingma, D.P., Ba, J.: Adam: a method for stochastic optimization. arXiv preprint arXiv:1412.6980 (2014)

Srivastava, N., Hinton, G., Krizhevsky, A., Sutskever, I., Salakhutdinov, R.: Dropout: a simple way to prevent neural networks from overfitting. J. Mach. Learn. Res. **15**(1), 1929–1958 (2014)

Optimizing Medical Image Classification Models for Edge Devices

Areeba Abid[1(✉)], Priyanshu Sinha[2], Aishwarya Harpale[3], Judy Gichoya[1], and Saptarshi Purkayastha[4]

[1] Emory University School of Medicine, Atlanta, GA, USA
areeba.abid@emory.edu
[2] Mentor Graphics India Pvt Ltd., Noida, India
[3] CakeSoft Technologies Pvt Ltd., Pune, Maharashtra, India
[4] Indiana University–Purdue University Indianapolis School of Informatics and Computing, Indianapolis, IN, USA
saptpurk@iupui.edu

Abstract. Machine learning algorithms for medical diagnostics often require resource-intensive environments to run, such as expensive cloud servers or high-end GPUs, making these models impractical for use in the field. We investigate the use of model quantization and GPU-acceleration for chest X-ray classification on edge devices. We employ 3 types of quantization (dynamic range, float-16, and full int8) which we tested on models trained on the Chest-XRay14 Dataset. We achieved a 2–4x reduction in model size, offset by small decreases in the mean AUC-ROC score of 0.0%–0.9%. On ARM architectures, integer quantization was shown to improve inference latency by up to 57%. However, we also observe significant increases in latency on x86 processors. GPU acceleration also improved inference latency, but this was outweighed by kernel launch overhead. We show that optimization of diagnostic models has the potential to expand their utility to day-to-day devices used by patients and healthcare workers; however, these improvements are context- and architecture-dependent and should be tested on the relevant devices before deployment in low-resource environments.

1 Background

1.1 Motivation

Machine learning algorithms in healthcare show promise for alleviating disparities in access to healthcare, by providing automated diagnostic support in low-resource areas. However, these models are often developed (and limited to being run on) high-end hardware or cloud servers. To achieve equity in machine learning access and take advantage of widespread mobile access in limited resource settings, these models should be tested on edge devices, rather than being limited to high-powered servers. There are advantages and limitations of edge devices

A. Abid and P. Sinha—Contributed equally.

© The Author(s), under exclusive license to Springer Nature Switzerland AG 2022
K. Matsui et al. (Eds.): DCAI 2021, LNNS 327, pp. 77–87, 2022.
https://doi.org/10.1007/978-3-030-86261-9_8

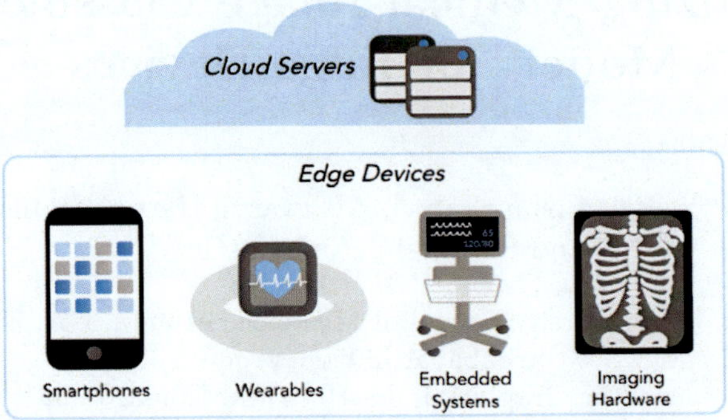

Fig. 1. Common machine learning contexts in healthcare.

and high powered server architectures. Cloud servers have considerably more computational capacity and memory, but are expensive. Network bandwidth is finite, so downloading models or uploading data to servers is a resource-heavy task. On the other hand, edge devices are constrained in terms of memory and computing power, but are cheaper and not limited by network speed, making them more accessible in environments where financial and/or network resources are limited. In medical use cases, edge devices are additionally advantageous because they are not bound by HIPAA restrictions that limit the transfer of electronic protected health information (eHPI) to the cloud, as patient information does not need to leave the device in order to obtain a model inference [1]. Models can also be deployed to wearables in the context of chronic disease management, such as predicting blood glucose levels [2]. Due to the memory, inference latency, and privacy advantages offered by edge devices, deep learning models in healthcare gain much utility when optimized for deployment and execution in lower-resource environments.

In this study, we evaluate the advantages and limitations of optimizing clinical machine learning models for edge devices. We study three metrics: model size, inference latency, and model accuracy as represented by area under the receiver operating characteristic (AUC-ROC) curve. We use radiology models trained on the NIH Chest-XRay14 Dataset and optimize these models by using 3 types of quantization: dynamic range, float-16, and full int8 quantization [3].

We hypothesize that if clinical models can be run faster and with reduced memory usage requirements on edge devices, models will be more suitable for deployment in limited-resource clinical settings for timely decision support.

1.2 Overview of Compression Techniques

Broadly, there are two common techniques for model compression for edge devices, *quantization* and *pruning* [6]. These methods reduce model complexity at a slight cost to accuracy, but offer improvements in computational speed and model size. This has the added benefit of reducing power consumption and bandwidth utilization.

Quantization is the reduction of model values to lower-bit representations. For example, a model trained with float32 precision can be compressed to use float16 or int8 precision for its weights, biases, and/or activations. This reduction in precision often has little impact on model accuracy, but reduces memory usage up to 4x (such as when reducing 32-bit floats to 8-bit integers). Computational speed also improves, as integer operations are generally faster than float operations on ARM CPUs [7]. Quantization can be implemented after model training (post-training quantization) or during training (quantization-aware training). Post-training quantization is faster to implement, but quantization-aware training typically offers better accuracy [8].

Connection Pruning is another common compression technique. In magnitude-based weight pruning, we drop low-weight connections between neural network layers. Models are represented by tensors; after connection pruning, the dropped connections are replaced by zeroes in the tensors. This results in sparse tensors for weight representations, which are easier to compress and reduce memory usage. As sparse representations contain zeroes, we can skip them while calculating an inference, reducing latency.

In this paper, we explore the advantages of quantization only, both post-training and quantization-aware training methods.

2 Method

2.1 Dataset

To demonstrate the efficacy of quantization for clinical use cases, we used the Chest-Xray14 Dataset, which consists of 112,120 X-ray images from 30,805 unique patients [3]. This dataset has been widely used to develop classification models for cardiopulmonary pathology. Each image in this dataset is annotated with labels from 14 pathology classes derived using text-mining from the associated radiology reports. The X-ray images can contain multiple pathologies, and each detected pathology is represented in a 1-by-14 vector as a positive class.

We randomly split the dataset into training (54,091 images), validation (23,183 images), and test (33,118 images) sets while ensuring that there was no patient overlap between each split. We performed pre-processing on each image by downscaling to 224 $\times$ 224 pixels.

2.2 Baseline FP32 Model

The floating point-32 model used as a baseline was Arevalo and Beltran's Chest X-Ray classification model ("Xrays-multi-dense121 0980aa"), developed using the DenseNet121 architecture [4]. This architecture consists of dense blocks followed by convolution and pooling layers. Each dense block receives feature maps from all the preceding layers and concatenates them to achieve a thinner and compact network. The model was initialized with weights pre-trained on the ImageNet dataset. An Adam optimizer is used to minimize the cost function starting with an initial learning rate of 0.001. Data generators were initialized with a batch size of 32.

Since each image can contain pathology in multiple classes, the output of the model is a 1 × 14 vector representing a probability score for each of the 14 pathology classes. The "No Finding" class is represented by a vector consisting of all zeroes.

We use this model as the baseline for size, inference latency, and accuracy comparison, and for generation of compressed models using post-training quantization.

2.3 Quantization of the Model

We implemented 3 types of quantization - Dynamic Range Post-Training Quantization, Float16 Post-Training Quantization, and Full Int8 Quantization-Aware Training (Table 1).

Table 1. Summary of model quantization techniques

Compression technique	Model architecture
Dynamic Range Post-Training Quantization	Only weights are reduced to 8-bit integers. Activations are dynamically down-scaled to 8-bit at run-time for faster computation
FP16 Post-Training Quantization	Weights, biases, activations, and outputs are all reduced to 16-bit floats
Full Int8 Quantization-Aware Training	Weights, biases, and activations are all reduced to 8-bit ints. Input & output remain 32-bit floats

These compression methods were implemented using TensorFlow Model Optimization Toolkit [5].

Dynamic Range Post-Training Quantization: In dynamic range quantization, the model combines both floating point and 8-bit integer precision. The model weights and biases are scaled down from floating point to integer precision statistically by calculating the scale factor and zero point beforehand. Activations are dynamically quantized at run-time to 8-bit precision, so that computations between activations and weights can be performed in 8-bit precision. However, the outputs are then converted back and stored as floats. This reduces the latency of computation at inference close to that of fixed-point integer operations. This typically reduces the model size by about 75%.

Float16 Post-Training Quantization: We implemented Float-16 Quantization to reduce the full precision 32-bit floating point data to reduced precision. The weights, biases and activations are stored in 16-bits, and computations occur in the same format. Outputs are also stored in the 16-bit format. This typically reduces the model size by about 50%.

Full Int8 Quantization-Aware Training: During training of the initial FP32 model, quantization is emulated by down-scaling float precision so it matches 8-bit precision. The initial FP32 model is trained with floating point values, but during the forward pass of information, these values are converted to int8 and then back to FP32. This makes the float values less precise during training, which makes the model robust to quantization. After training, the model weights are quantized to int8, but since the weights were determined at lower precision, there is no additional loss of accuracy from the less-precise weights. Biases and activations are also reduced to int8, which does have some cost to accuracy, but allows for faster computation. This typically reduces the model size by about 75%.

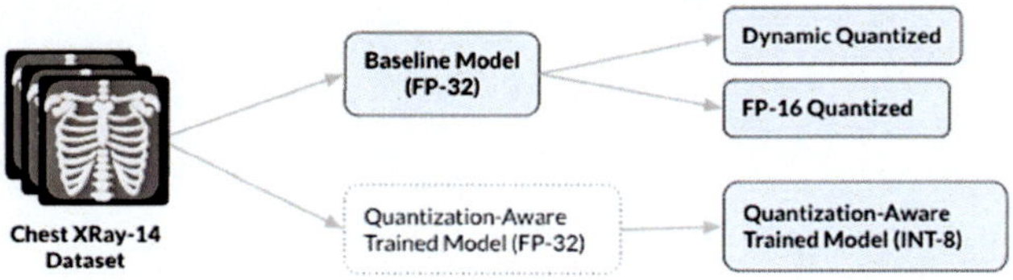

Fig. 2. Development of quantized models.

2.4 Hardware Specifications and Costs

Edge devices come in a variety of hardware specifications, ranging from smartphones to an array of embedded devices and chips. To obtain representative results, we tested our compression techniques on a range of ARM and x86 architectures. The ARM devices used are the NVIDIA Jetson Nano, Raspberry Pi 3B+, Google Pixel, and Samsung Galaxy S10+. The x86 devices used are PC laptops using Intel x86 and AMD x86 processors. The costs of these devices range from less than 50 USD to over 1000 USD (Table 2). These costs can be compared to the costs of typical GPUs, both cloud and local (Table 3).

The choice of which devices to use was made based on CPU architecture, price range, and GPU availability. We include both ARM and x86 processors to investigate the effect of quantization on different architectures, which is critical to measure because these processors are optimized for computations on different data types: Arm processors have Integer computation accelerators, whereas x86 processors have floating point accelerators. We also examine the performance of the models over a spectrum of high-end and low-end devices, since a major application of edge computing is in resource-constrained environments. We include smartphone devices with optional GPU enabling, to allow for comparison of CPU- vs GPU- inference on a single device.

2.5 Measuring Accuracy and Inference Latency

To measure the effect of model compression on *accuracy*, 33,118 test images from the ChestX-Ray 14 Dataset were used to evaluate the models and obtain AUC-ROC curves. For *inference latency*, each model was tested with 25 distinct preprocessed Chest X-Ray images. The model file was closed and reloaded between

Table 2. Specifications of test devices: architecture and cost.

Device name	ARM/x86	Specification	Price range
NVIDIA Jetson Nano	ARM	ARMv8 Processor rev 1 (v8l), 4 GB RAM, 4 Core CPU	<100 USD
Raspberry Pi 3B+	ARM	4 Core CPU, 1 GB RAM, 1.4 GHz processor, Broadcom Arm Cortex A53-architecture processor	<50 USD
Google Pixel (1st Gen)	ARM	Quad-core (2×2.15 GHz & 2×1.6 GHz) Kryo 64-bit ARMv8-A	650 USD
Samsung Galaxy S10+	ARM	CPU: Snapdragon 855; GPU: Adreno 640	999 USD
PC Laptop, Intel processor	x86	Intel Core i7-7820HQ CPU 2.90 GHz, 32 GB RAM, 8 Core	>1000 USD
PC Laptop, AMD processor	x86	Ryzen 7 3750H, 16 GB, 8 Core. Nvidia GeForce GTX 1660Ti with Max Q design GPU. VRAM: 6 GB	>1000 USD

Table 3. Costs of representative GPU servers (cloud and local examples)

Type of server	Host/Service	Cost
Local GPU	ThinkStation Nvidia GeForce RTX2080 Super 8 GB GDDR6 Graphics Card	1,100 USD [9]
Local GPU	Dell 16 GB NVIDIA Tesla T4 GPU Graphic Card	3,904 USD [10]
Cloud GPU	Amazon Machine Learning	0.42 USD/h + 0.10 USD per 1000 predictions [11]
Cloud GPU	Google Cloud, Basic-GPU	0.83 USD/h [12]
Cloud GPU	Microsoft Azure, NC6 GPU	0.90 USD/h [13]

each image test. Two measurements were recorded; the total run time of the test for all 25 images, and the average inference latency (not including model and image file load time). The first measurement takes into account model load time and GPU kernel creation (if applicable), while the second measurement isolates inference latency only.

2.6 Code Repository

The code used to evaluate models is located at this repository: https://github.com/areeba-a-abid/OptimizationEdgeDevices.

3 Results and Discussion

The compression techniques used in this paper demonstrate varying degrees of improvement in model metrics for each device tested. The tables below describe the impact of model compression techniques on model size, model accuracy, and inference time.

Table 4. Model accuracy (AUROC) by class. Note: differences >0.05 between optimized models and baseline are bolded.

Class	Baseline FP32	Dynamic quantization	Float16 quantization	Quant-aware trained Int8
Atelectasis	0.78	0.78	0.78	0.75
Cardiomegaly	0.90	0.90	0.90	**0.71***
Consolidation	0.79	0.79	0.79	0.77
Edema	0.88	0.88	0.88	**0.79***
Effusion	0.87	0.87	0.87	0.84
Emphysema	0.88	0.88	0.88	**0.78***
Fibrosis	0.79	0.79	0.79	**0.73***
Hernia	0.83	0.83	0.83	**0.51***
Infiltration	0.71	0.70	0.71	**0.66***
Mass	0.82	0.82	0.82	**0.75***
Nodule	0.73	0.73	0.73	**0.65***
Pleural Thickening	0.77	0.77	0.77	**0.71***
Pneumonia	0.74	0.74	0.74	**0.66***
Pneumothorax	0.85	0.85	0.85	**0.77***
Mean AUC-ROC	0.81	0.81	0.81	**0.72***

Table 5. Size of models

Architecture	Baseline model	Dynamic quant.	FP16 quant.	QAT Int8
Model size (MB)	27.9	7.3	14.1	7.4
Size reduction	–	$\tilde{3}$.8x	$\tilde{2}$.0x	$\tilde{3}$.8x

3.1 Model Accuracy

Our results demonstrate that compression of deep radiology models is possible with very minor cost with respect to accuracy (Table 4). The Dynamically Quantized and FP16 Quantized models performed almost identically to the baseline model that they were derived from, with mean AUC-ROCs of 0.81 for both (the same as the mean AUC-ROC of the baseline FP32 model). The QAT Full Int8 model observed a decrease of 0.09 in mean AUC-ROC, with the 'Hernia' and 'Cardiomegaly' classes experiencing the biggest drop. In medical contexts, a class-by-class comparison is necessary to investigate where losses are most pronounced and for which types patients inferences should be used with extra caution. A reduction in sensitivity for high-mortality or costly conditions may lead to a larger impact on model utility as compared to the same percent reduction in the labeling-accuracy of benign conditions.

3.2 Model Size

The baseline FP32 model, which used 32-bit float representation for weights and activations, had a model size of 27.9 MB. FP16 Quantization reduced size by almost half, to 14.1 MB. By reducing representations to 8 bits, Dynamic and Int8 Quantization offered almost a 4x reduction, to 7.3 and 7.4 MB (Table 5).

3.3 Inference Latency

The degree of improvement in inference time was architecture-dependent. On ARM devices, integer computations are faster, so latency improved most with Dynamic Quantization and QAT Int8 Quantization. However, the more expensive smartphones (Google Pixel & Samsung Galaxy) did not show this improvement, because the Snapdragon CPUs do not support all quantized operations; investigating this is beyond the scope of this paper. The reduction in latency was most significant for the cheapest device (the Raspberry Pi), improving by 49–57% for these two integer quantization methods. The Jetson Nano and Samsung Galaxy demonstrated improvements between 25–49% (Table 6). FP16 quantization also offered a 13% reduction in latency on the Raspberry Pi, but increased latency on all other ARM devices. No significant improvement of FP16 models on ARM devices was expected, because the computations are still conducted using floats. The percent change for each model on ARM devices is shown in Fig. 3.

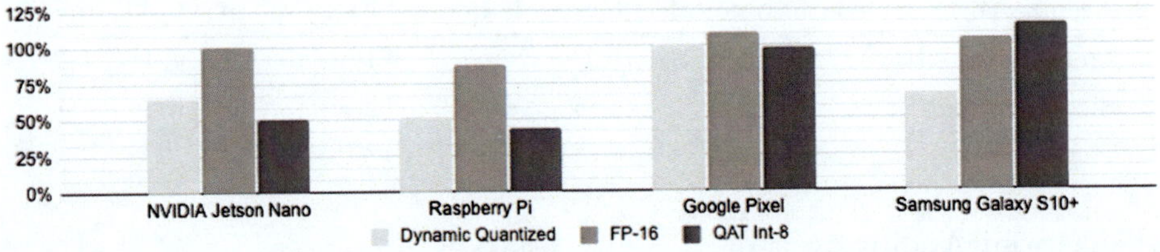

Fig. 3. Percent change in inference latency for ARM devices compared to baseline.

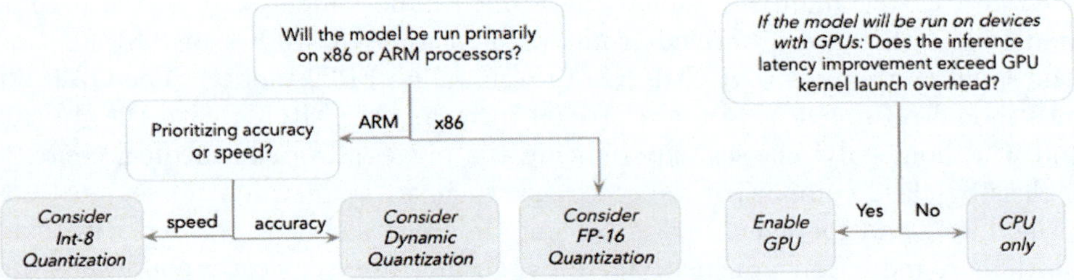

Fig. 4. To optimize or not to optimize? Factoring in device types and priorities in optimization decisions.

Table 6. Inference latency (ms) per image and percent change from baseline per device

Architecture	Baseline model	Dynamic quant.	FP16 quant.	QAT Int8
NVIDIA Jetson Nano	801	520 (↓35%)	806 (↑1%)	410 (↓49%)
Raspberry Pi 3B+	2340	1200 (↓49%)	2037 (↓13%)	1010 (↓57%)
Google Pixel	684	690 (↑1%)	741 (↑8%)	674 (↓1%)
Samsung Galaxy S10+	191	128 (↓33%)	201(↑5%)	220 (↑15%)
Intel x86	50	6070 (↑12040%)	50 (0%)	6937 (↑13774%)
AMD x86	100	5980 (↑5880%)	89 (↓11%)	6305 (↑6205%)

For x86 devices, quantization methods that convert weights to integers actually increase latency by over 100x on the Intel processor and over 50x on the AMD processor. This is expected, as x86 devices are optimized for float computations. While integer quantization offers improvements for ARM devices, the dramatic effect on latency for x86 processors is a significant drawback to consider.

Investigating the effect of GPU on latency was done using the Samsung Galaxy S10+. When GPU is enabled, the time for inference per image is reduced for all models, but the overall run-time of the prediction increases (Table 7).

Table 7. Effect of GPU on inference latency and total run time (ms) on Samsung Galaxy S10+

	Architecture	Baseline model	Dynamic quant.	FP16 quant.	QAT Int8
Inference time only (per image)	CPU only:	191	128	201	220
	GPU-enabled:	124	125	120	124
Average run time (per image)	CPU only:	195	132	206	223
	GPU-enabled:	981	990	1002	1239

This is because setup time for the device's GPU kernel is expensive. Whether this trade-off is worthwhile is dependent on the number of images being passed into the model; beyond a certain number of inferences, the speedup of GPU surpasses the initial cost of setup.

The decision on if and how to optimize for edge devices is outlined in Fig. 4.

4 Conclusion

We find that model compression is an effective way to reduce model size by 2–4x with a minimal reduction in accuracy. This allows for a significant reduction in device cost and makes clinical models more accessible for a wider range of patients and healthcare providers, especially as machine learning models expand to a wide range of edge devices, such as smartphones, wearable technology, embedded devices, and imaging hardware. However, given the diversity of devices used in medicine, it is important to note that the impact of model compression on inference latency varies depending on the architecture. Because x86 processors are optimized for float calculations, quantization to integers increases latency. Therefore, integer quantization methods are best suited for devices using ARM architectures.

Improvements in latency for x86 processors are demonstrated using FP16 models. Enabling GPU on higher end devices that have this option can also improve performance, but has the added cost of GPU kernel setup time. For example, in the context of radiology, a model that reads a single patient's X-ray images on-demand may be better off not utilizing GPU optimizations, but a use-case in which many patients' images are read at once may benefit from it.

As the availability of medical machine learning grows, we show that careful choices about model compression allow these advancements to be made more widely accessible, independent of access to high-cost devices or servers, but that the improvements offered by quantization are architecture- and context-dependent.

5 Future Work

In this paper, we used image classification as an example of the types of clinical models run on edge devices. Optimization for edge devices should also be explored in image segmentation problems, such as automated ultrasound segmentation; this is a rapidly growing area of model development and can be used at point-of-care on edge devices [14].

There are many more methods of model compression that should be further investigated. Quantization can be used to reduce model size further by reducing precision to 4-, 2-, or even 1-bit. Connection pruning, as discussed in Sect. 1.2 can offer improvements independent of the type of processor used, and should be further explored.

Power usage is another important metric that was not explicitly investigated in this paper. Compression is expected to reduce the power usage of a model as a function of reduced run-time, but measuring and confirming this assumption was beyond the scope of this paper.

References

1. U.S. Department of Health and Human Services: Guidance on HIPAA and Cloud Computing (2020). https://bit.ly/3wyHFxD
2. Bhimireddy, A., et al.: Blood glucose level prediction as time-series modeling using sequence-to-sequence neural networks. In: CEUR Workshop Proceedings (2020)
3. Wang, X., Peng, Y., Lu, L., Lu, Z., Bagheri, M., Summers, R.M.: ChestX-Ray8: hospital-scale chest x-ray database and benchmarks on weakly-supervised classification and localization of common thorax diseases. In: IEEE Conference on Computer Vision and Pattern Recognition (2017). https://arxiv.org/pdf/1705.02315v5.pdf
4. Arevalo, W., Beltran, J.: Xrays-multi-dense121 0980aa (2019). https://www.kaggle.com/willarevalo/xrays-multi-densenet121/data?select=weights.h5
5. TensorFlow. Optimize machine learning models. https://www.tensorflow.org/model_optimization
6. Massimo, M., et al.: Edge machine learning for AI-enabled IoT devices: a review. Sensors **20**(9), 2533 (2020)
7. Ilin, D., et al.: Fast integer approximations in convolutional neural networks using layer-by-layer training. In: Ninth International Conference on Machine Vision (ICMV 2016), vol. 10341. International Society for Optics and Photonics (2017)
8. Tensorflow. Quantization aware training. https://www.tensorflow.org/model_optimization/guide/quantization/training
9. ThinkStation Nvidia GeForce RTX2080 Super 8GB GDDR6 Graphics Card. https://lnv.gy/3fJvQhj
10. Dell 16GB NVIDIA Tesla T4 GPU Graphic Card. https://dell.to/3fiw14m
11. AWS: Build a Machine Learning Model. https://aws.amazon.com/getting-started/projects/build-machine-learning-model/services-costs/
12. Google Cloud AI Platform Pricing. https://cloud.google.com/ai-platform/training/pricing
13. Azure Machine Learning pricing. https://azure.microsoft.com/en-us/pricing/details/machine-learning/
14. AskariHemmat, M., et al.: U-Net fixed-point quantization for medical image segmentation. In: MICCAI's Hardware Aware Learning for Medical Imaging and Computer Assisted Intervention (2019). https://arxiv.org/abs/1908.01073

Check for updates

Song Recommender System Based on Emotional Aspects and Social Relations

Carlos J. Gomes, Ana B. Gil-González(✉), Ana Luis-Reboredo, Diego Sánchez-Moreno, and María N. Moreno-García

Department of Computer Science and Automation, Sciences Faculty, University of Salamanca, Plaza de los Caídos s/n, 37007 Salamanca, Spain
{cgomes,abg,adeluis,sanchez91,mmg}@usal.es

Abstract. Music streaming services have opened the possibility of accessing huge quantities of songs and more sophisticated data can be utilized by recommender systems to improve their performance. Some recommendation methods dealing with different music features have been proposed during the last years, but most of them do not consider emotional aspects. The recommender system presented in this work, allows the classification of the music into emotions from acoustic characteristics extracted directly by means of an automatic analysis of the songs. These emotional aspects of the songs are incorporated to the proposed recommendation models, that also include recommendations to groups of users from their social relationships. The experiments show an improvement in the recommendation reliability obtained by this proposal against the classic collaborative filtering recommendation approaches, for both individual and group recommendations.

Keywords: Recommender systems · Classification of emotions · Collaborative filtering · Music recommendation · Group recommendation

1 Introduction

Technological advances have allowed people to be more and more connected to music. Nowadays, music streaming services provide access to millions of songs giving users the opportunity to find their favorite music, as well as to explore and discover new musical contents fitting their preferences through the recommendation mechanisms that these platforms are endowed with. There are numerous proposals of recommendation methods but most of them can be classified into two main groups, content-based and collaborative filtering (CF). The former use properties of items to find similarities between them and recommend to a user items that are similar to those that he consumed or liked in the past. CF bases the recommendations of an item to a given user on evaluations about other items made by users with similar preferences. CF approach can also be classified in user-based and item-based methods. User-based techniques compute the similarity between users to make recommendations. In item-based CF similarity between items, obtained from the ratings received from users, are used instead of similarity between users. In addition to the described categories of methods, many hybrid approaches have

© The Author(s), under exclusive license to Springer Nature Switzerland AG 2022
K. Matsui et al. (Eds.): DCAI 2021, LNNS 327, pp. 88–97, 2022.
https://doi.org/10.1007/978-3-030-86261-9_9

been reported in the literature, aiming at improving recommendation reliability as well as benefiting from the advantages and avoiding the drawbacks characteristic of each group of methods.

Emotional aspects have been little exploited in recommender systems, despite the great interest they have aroused nowadays in the context of social networks where sentiment analysis is the focus of many studies. The connection between feelings or emotions and music has been proved in several researches where the relationship between different emotions and acoustic characteristics such as timbre, rhythm, intensity, mode, tonality, tempo (Beats per Minute - BPM) is checked [1]. In some of them the valence and activation scale of SAM (Self-Assessment Manikin) [2] is used to assess musical extracts and assign them an emotion corresponding to Russell's Circumplex Model [3].

In this work, these findings are exploited by means of a system that considers emotions aiming at providing users with more personalized music recommendations. The system also takes advantage of social relationships between users and implements a group recommendation method. The proposal is a hybrid approach that includes characteristics of both content-based and CF methods. It makes use of emotional aspects considering features of both songs and users and considering ratings given by users to songs to apply a CF technique. The system provides recommendations based on different aspects. These can be acoustic characteristics of the songs, emotions associated with the intensity of some of these characteristics and social relations between users.

2 Related Work

In the field of music, many recommender systems have been developed, especially since the popularization of music streaming services. Recent CF-based proposals in the literature for music recommendation are focused on avoiding usual problems of recommender systems such as sparsity and gray sheep problems [4] and improving the results of traditional methods. Content-based methods have also been widely applied in the music domain, in many cases to solve CF problems. In this context, content information can be provided by both music metadata (title, artist, genre, lyrics...) and audio features (timbre, melody, rhythm, harmony...) [5]. The works described above have contributed to improve some aspects of music recommender systems, however, the current trend is to resource to hybrid strategies to exploit the best part of each technique and avoid their problems [6].

Although emotions have not been extensively studied in the domain of recommender systems, the specific area of the music can be a very propitious field for exploiting emotions. In [7], recommendations based on the genres of the songs were tested, using just CF and CF with an emotion filter, achieving better results the second approach. A hybrid approach presented in [8] makes use of a weighting system based on user listening behavior to combine three different methods: content-based, CF and an emotion-based procedure that find interesting music for users from the differences between their interests and musical emotions. In [9] the mood is used as a context factor for music recommendations. The work presented in [10] considers artist listening habits of users in a hybrid mood-based system for artist recommendation. It involves a content-based approach in which similarity between artists is obtained from the acoustic features of their 10 most popular songs.

3 Methods

The aim of this work is to take advantage of emotional aspects to improve traditional CF recommender systems. Thus, we present a complete and operative proposal. Starting from an architecture that supports an application with streaming music loading service as well as the generation of a dataset and the validation of different proposals including recommendations to groups in the music social network.

3.1 Architecture

Figure 1 shows the client-server architecture of the web system in charge of storing songs, extracting acoustic characteristics, and classifying them in emotions. It makes use of Spotify services through its API by means of HTTP requests. The system is also endowed with a recommendation module that is responsible for predicting user preferences and providing different types of personalized suggestions of songs.

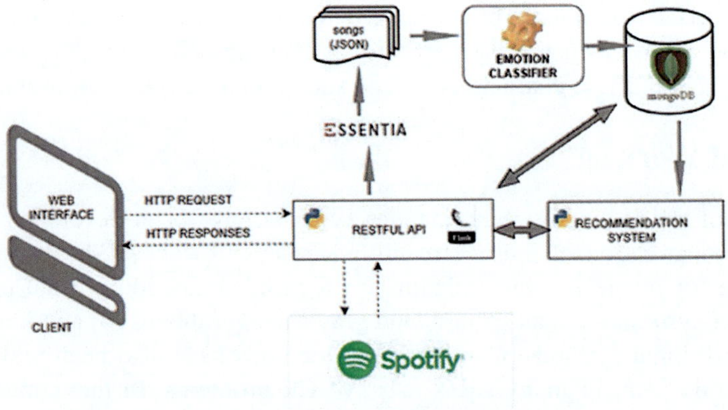

Fig. 1. MoodSically system architecture.

On this architecture we developed the application called MoodSically [11] that allows to create a personal music content manager for automatic detection of emotions, using acoustic features extracted from songs uploaded by users. This initial functionality has been improved by integrating automatic song analysis with the Spotify music streaming services, as well as introducing a module implementing a recommender system that makes use of those emotional aspects and/or acoustic features. In addition, the system, implemented as a web application, has some social network utilities as option of following between users. The analysis and extraction of acoustic features is made by means of the Essentia [12] and Gaia libraries (https://github.com/MTG/gaia). Those features are low-level descriptors such as mode, dance capacity, beats per minute (BPM), or volume; and high-level descriptors such as musical genre, timbre, tonality in order to build the probability of types of emotions (sad, happy, active or calm).

3.2 Classifier of Emotions

The classification of the songs by emotions is performed using values of valence and arousal that are obtained from values of probability of sad, happy, active, and calm extracted as high-level descriptors.

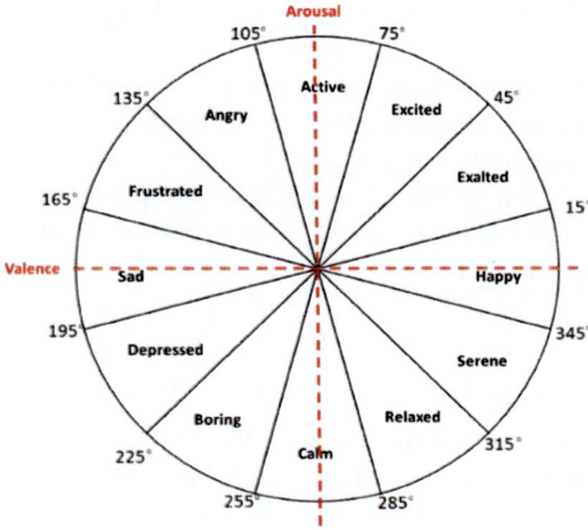

Fig. 2. Russell's 12-segment circumplex

The valence (Eq. 1) is calculated by subtracting the probability that the song is sad (*PSad*) from the probability that it is happy (*PHappy*). Arousal (Eq. 2) is obtained by subtracting the probability that the song is Calm (*PCalm*) from the probability that it is a song Active (*PActive*).

$$valence = PSad - PHappy \tag{1}$$

$$arousal = PCalm - PActive \tag{2}$$

These two variables are used to compute the polar coordinates, which are given by the angle (Eq. 3) and the value of the distance (Eq. 4).

$$\theta = atan2\left({}^{valence}/_{arousal}\right) \tag{3}$$

$$r = \sqrt{valence^2 + arousal^2} \tag{4}$$

The songs are classified by any of 12-Point Affect Circumplex (12-PAC) model of Core Affect, [13] (Fig. 2): Frustrated, relaxed, calm, excited, exalted, serene, boring, depressed, active, sad, happy and angry.

3.3 Song Recommendation Methodology

The proposed methodology for recommending songs is a hybrid approach combining CF and content-based methods. Besides ratings, attributes of songs and users were used to obtain similarities both between items and users by means of the cosine measure. This metric is specially indicated for problems in which more than one attribute is involved, as in this work, while Pearson coefficient is indicated when similarity is computed only from ratings.

User-based CF recommendations were obtained by applying the k-nearest neighbors (k-NN) algorithm. The aim of this method is to find the k users (neighbors) with the highest coefficient of similarity to the active user (using both explicit ratings and additional attributes). Once the k neighbors are obtained, the prediction of the rating that the active user would give to a given song *i* is computed from the ratings given by his/her neighbors to this song (Eq. 5).

$$P_{ai} = \overline{r_a} + \frac{\sum_{u \in K} (r_{ui} - \overline{r_u}) \times w_{ai}}{\sum_{u \in K} w_{au}} \tag{5}$$

Where P_{ai} is the prediction of the rating for the active user and the song$i$, w_{au} is the similarity between the active user and the user$u$, K is the subset of k similar users, r_{ui} is the rating that user u gives to item i and $\overline{r_u}$ is the average of ratings for user u.

Item-based CF approach uses the transposed matrix (items x users) for predictions. Predictions for active users in this approach were computed by using Eq. 6, which considers similarity between items according to the ratings given to them by users. Where K is the set of k items that are the most similar to item i and w_{ij} is the similarity between items i and j.

$$P_{ai} = \frac{\sum_{j \in K} \left(r_{aj} \times w_{ij} \right)}{\sum_{j \in K} w_{ij}} \tag{6}$$

3.4 Recommendations to Groups

The internet produces a large amount of rich and complex data, which allow to identify varied information of the users, enabling inference about their interests. Recent works such as [14, 15] have proved that incorporating social information into recommendation models can improve recommendation performance. However, this aspect has been lesser studied in the music domain, given the lack of social data in these systems.

The web application implemented in this work acts as a social network endowed with options of user following, which allows to take advantage of that information for making recommendations to groups. The technique used for group recommendations involves the predictions of ratings for the active user and the users he/she follows. The Average Satisfaction Maximization strategy (Eq. 7) is used, where the group rating is calculated as the average of the individual ratings.

$$R_i = \frac{1}{n} \sum_{u \in G} r_{ui} \tag{7}$$

Where R_i is the rating for the group, u represents each user of the group G, r_{ui} is the predicted rating of user u about the item i and n is the number of users belonging to the group G.

4 Results

This section presents the results obtained in the validation of the recommender system, for both individual and group recommendations.

4.1 Evaluation Dataset

The dataset used contains information from 20 users and 1191 songs. Users have 1306 interactions with the application for rating songs and 28 user-user user relationships in the social network.

Due to limitations established by the streaming service, it is only possible to extract 30 s of each song for acoustic analysis. To find out the impact of using that portion of time instead of the entire song, differences in the classification of a sample of songs for both cases were analyzed. For most of them, the classification was the same, although little differences took place because of variations in the values of both the high- and low-level attributes used to obtain valence and activation values. Since these discrepancies were reflected as classification into adjacent emotions using Russell's 12-segment Circumplex (Fig. 2), we consider that they can be assumed because their impact on outcome is not significant. The same assumption was made at the 2007 Annual Music Information Retrieval and Evaluation Conference [16] achieving good results.

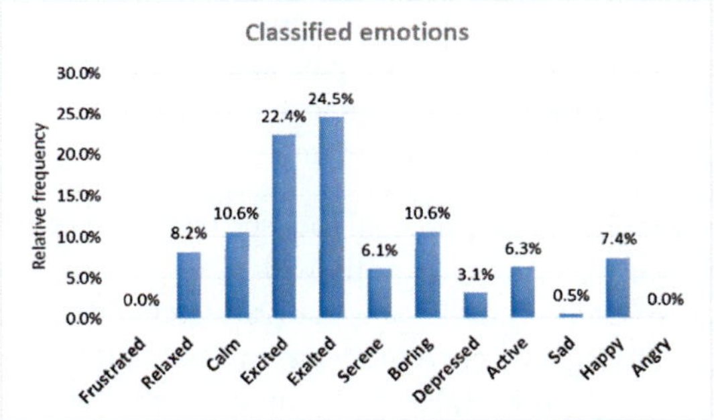

Fig. 3. Frequency of emotions in the dataset

Regarding the emotions that are automatically generated and associated with the songs in the dataset, we can see in Fig. 3 that there is a great variety.

The application stores a score from 1 to 5 that indicates the level of compliance of users with the emotion generated by the system using a rating scale from 1 to 5 (representation by stars) (Fig. 4). In general terms, there is a good association between emotions extracted and songs according to the perspective of users. Figure 4 also shows an evaluation mechanism for songs using emoticons. This system allows user to establish a preference degree for the last 50 listened songs through Spotify. This score will be used as the rating needed to apply CF (Eqs. 5 and 6).

Fig. 4. System for emotion validation and evaluation of songs

As supplementary information to be used by the recommender system, the number of songs associated with each emotion played by each user is stored as a user attribute. In addition, descriptors of both low and high level are extracted from the songs. These are: bpm, dissonance, emotions generated by the system, emotions probability (happy, active, calm and sad), timbre, tonal and volume.

4.2 Experiments

In all experiments, 10-fold cross validation was applied to test the reliability of the methods. We utilize the indicator RMSE (Root Mean Squared Error) as the metric used in the evaluation to compare the recommendation performance between different models. This measure quantifies the difference between the predicted rating and the actual value. For this rating system, RMSE can vary in the range [0, 4].

The performance of the methods implemented in the recommender system based on emotional aspects was carried out by comparing their results with classical CF approaches. Baseline methods used to compare our proposals where both user-based and item-based CF. Several values of k were tested when applying the k-NN algorithm to find similarity between users and between items.

User-Based CF Approach. In our proposal for user-based CF approach, emotions from songs played by users were considered jointly with users' ratings in order to find similar users. The percentage of each of the 12 automatically generated emotions is calculated from the total number of songs listened to by each user, adding a new attribute for each emotion. The results are given in Table 1. It can be seen that the introduction of the emotional aspects results in a decrease in the error rate for all values of k.

Table 1. RMSE for user-based CF

Method	k = 5	k = 10	k = 15
K-NN	0.9053	0.9046	0.9046
K-NN with emotions	**0.8994**	**0.8899**	**0.8807**

Item-Based CF Approach. In the following tests, errors for different proposals were analyzed considering the item-based approach. They are described below taking as reference the base experiment item-based k-NN:

- Proposal 1: Besides rating, a new attribute is considered to compute similarity between items. It is the probability that a song causes one of these four emotions: happy, sad, active, and calm. These probabilities are obtained as high-level descriptors in acoustic analysis.
- Proposal 2: Emotion probabilities are also considered, and an additional attribute: the arithmetic mean of user evaluations of the emotions of each of the songs.
- Proposal 3: Starting from experiment 2 but considering only the average of the evaluations of emotions.
- Proposal 4: Similarity is computed from ratings and some of the acoustic characteristics that most influence the determination of emotional aspects (volume, dissonance, timbre, bpm and tone).

Table 2. RMSE for the experiments conducted using item-based CF

Method	k = 10	k = 50	k = 100	k = 200
K-NN	1.0774	1.0195	1.0166	1.0166
Proposal 1	0.8251	0.9191	0.9403	0.9608
Proposal 2	0.8395	0.8981	0.9135	0.9400
Proposal 3	0.9542	0.9151	0.9423	0.9429
Proposal 4	**0.7157**	**0.7992**	**0.8461**	**0.8922**

The results of the error, as seen in Table 2, have also improved slightly in this case with respect to the base approach. These data show us the proposal 4 reaches the best results, using as attributes the set of acoustic characteristics extracted as high-level descriptors. An improvement of almost 8% has been achieved for k = 10.

Recommendations to Groups. Social information of the system has been used to make predictions using the set of users that an active user follows. In this case, recommendations are made for the active user and for all the users he has chosen to follow. In this case we have compared the results of the recommendations for groups without emotions and with emotions.

Table 3. RMSE for user-based CF and recommendations to groups

Approach	k = 5	k = 10	k = 15
Groups	0.7982	0.7804	0.7831
Groups with emotions	**0.7865**	**0.7128**	**0.7045**

Recommendations to groups considering only explicit user ratings are made using user-based CF for individual recommendations and the aggregation method of maximizing the average satisfaction described in Sect. 3.4. For groups with emotions the same approach was used but considering the percentage of emotions associated with the songs. The results in Table 3 show that the error decreases by up to 10% using the approach of recommendations to groups based on emotions for k = 15, which is a very significant result.

5 Conclusions

In conclusion, a web application has been implemented that allows automatic extraction of acoustic characteristics of the songs and their classification by emotions, all using information from the songs played by users and provided by the Spotify API. In addition, a recommendation module has been implemented making use of the two-basic CF approaches (User-based and item-based). These approaches were enhanced using hybrid methods and attributes of both acoustic characteristics and emotions.

In view of the results, we can conclude that the item-based approach provides better results than the user-based approach, especially when using attributes of musical characteristics such as bpm, tone, volume or dissonance, and small values of the number of neighbors. The good results obtained in the recommendations to groups of users using probabilities of emotions are interesting, with an improvement of almost 10% in one of the cases under study. In future work, we will consider constructing a more rational and sophisticated rating prediction function for recommendation through the incorporation of context-Aware Recommender Systems (CARS).

Acknowledgements. This research has been supported by the Department of Education of the Junta de Castilla y León, Spain, (ORDEN EDU/667/2019). Project code: SA064G19.

References

1. Kawakami, A., Furukawa, K., Katahira, K., Kamiyama, K., Okanoya, K.: Relations between musical structures and perceived and felt emotions. Music Percept. Interdiscipl. J. **30**(4), 407–417 (2012)
2. Bradley, M.M., Lang, P.J.: Measuring emotion: The self-assessment manikin and the semantic differential. J. Behav. Ther. Exp. Psychiatry **25**, 49–59 (1994)
3. Russell, J.: A circumplex model of affect. J. Personal. Soc. Psychol. **39**(12), 1161–1178 (1980)
4. Sánchez-Moreno, D., Gil, A.B., Muñoz, M.D., López, V.F., Moreno, M.N.: A collaborative filtering method for music recommendation using playing coefficients for artists and users. Expert Syst. Appl. **66**, 234–244 (2016)
5. Kuo, F.F., Shan, M.K.: A personalized music filtering system based on melody style classification. In: Proceedings of the IEEE International Conference on Data Mining, pp. 649–652 (2002)
6. Yoshii, K., Goto, M., Komatani, K., Ogata, T., Okuno, H.G. : Hybrid collaborative and content-based music recommendation using probabilistic model with latent user preferences. In: Proceedings of the 7th International Conference on Music Information Retrieval, pp. 296–301 (2006)

7. Mortensen, M., Gurrin, C., Johansen, D.: Real-world mood-based music re-commendation. Inf. Retrieval Technol. AIRS **2008**, 514–519 (2008)
8. Lu, C.C., Tseng, V.S.: A novel method for personalized music recommendation. Expert Syst. Appl. **36**, 10035–10044 (2009)
9. Baltrunas, L., et al.: InCarMusic: context-aware music recommendations in a car. In: Huemer, C., Setzer, T. (eds.) EC-Web 2011. LNBIP, vol. 85, pp. 89–100. Springer, Heidelberg (2011). https://doi.org/10.1007/978-3-642-23014-1_8
10. Andjelkovic, I., Parra, D., O'Donovan, J.: Moodplay: interactive mood-based music discovery and recommendation. In Conference: User Modeling, Adaptation and Personalization, At Halifax (2016)
11. Vicente, G., Gil, A.B., de Luis Reboredo, A., Sánchez-Moreno, D., Moreno-García, M.N.: Moodsically. personal music management tool with automatic classification of emotions. In International Symposium on Distributed Computing and Artificial Intelligence, pp. 112–119. Springer, Cham (2018)
12. Bogdanov, D., et al.: Essentia: an open-source library for sound and music analysis. In: Proceedings - 21st ACM International Conference on Multimedia (2013)
13. Yik, M., Russell, J.A., Steiger, J.H.: A 12-point circumplex structure of core affect. Emotion **11**(4), 705 (2011)
14. Sánchez-Moreno, D., Pérez-Marcos, J., Gil, A.B., López, V.F., Moreno-García, M.N.: Social influence-based similarity measures for user-user collaborative filtering applied to music recommendation. Adv. Intell. Syst. Comput. **801**, 1–8 (2019)
15. Pérez-Marcos, J., Martín-Gómez, L., Jiménez-Bravo, D.M., López, V.F., Moreno-García, M.N.: Hybrid system for video game recommendation based on implicit ratings and social networks. J. Ambient. Intell. Humaniz. Comput. **11**(11), 4525–4535 (2020). https://doi.org/10.1007/s12652-020-01681-0
16. Hu, X., Downie, J., Laurier, C., Bay, M., Ehmann, A.F.: The 2007 mirex audio mood classification task: lessons learned. In: ISMIR 2008 - 9th International Conference on Music Information Retrieval, pp. 462–467 (2008)

Check for updates

Non-isomorphic CNF Generation

Paolo Fantozzi[1](✉), Luigi Laura[2], Umberto Nanni[1], and Alessandro Villa[2]

[1] Sapienza University of Rome, Rome, Italy
paolo.fantozzi@uniroma1.it, nanni@diag.uniroma1.it

[2] International Telematic University Uninettuno, Rome, Italy
luigi.laura@uninettunouniversity.net,
a.villa2@students.uninettunouniversity.net

Abstract. The Graph Isomorphism (GI) Class is the class of all the problems equivalent to the Graph Isomorphism problem, that is not known to be solvable in polynomial time nor to be NP-complete. GI thus is a very interesting complexity class that may be in NP-intermediate. In this work we focus on the CNF Syntactic Formula Isomorphism (CSFI) problem, that has been proved to be GI-complete, and we present a formal approach to the definition of "trivial non-isomorphic" instances and an algorithm to generate "non-trivial" instances. The applications of such generator are twofold: on the one side we can use it to compare deterministic algorithms, and on the other side, following recent approaches for NP-complete problems such as SAT and TSP, we can also use the generated instances to train neural networks.

Keywords: NP complete problems · Graph Isomorphism · CNF Syntactic Formula Isomorphism

1 Introduction

In the quest for the P versus NP problem an important role might be played by the Graph Isomorphism (GI) Class, a candidate to be in NP-intermediate if $P \neq NP$. The GI class includes the eponymous Graph Isomorphism problem and other graph related problems; in recent years other problems have been shown to be in GI, such as CNF Syntactic Formula Isomorphism (CSFI) problem [3], that is the problem, given two Conjunctive Normal Form (CNF) formulas, to decide whether there is a permutation of the clauses and the literals such that the two formulas are the same one.

Given its similarity, at least from a formal point of view, with the SAT problem, the CSFI problem is very interesting to study. Indeed, in many different classes of complexity, there are problems that are composed by different kind of instances. Some of them are very difficult to solve, while some others are very easy to solve. This is the case of the CSFI.

Furthermore, the Formula Isomorphism problem (that is more general than CSFI) is in the second level of the polynomial hierarchy: $\Sigma_2\mathbf{P}$, but also it cannot

© The Author(s), under exclusive license to Springer Nature Switzerland AG 2022
K. Matsui et al. (Eds.): DCAI 2021, LNNS 327, pp. 98–107, 2022.
https://doi.org/10.1007/978-3-030-86261-9_10

be $\Sigma_2\mathbf{P}$-complete unless the polynomial time hierarchy collapses, as studied by Agrawal and Thierauf and shown in [1,2]. Also, in [10], it is shown that Graph Isomorphism (GI) is polynomial time reducible to FI.

Since that CSFI is part of that complexity class, that it means that there should exist some instances that could be solved in polynomial time and some others that cannot be solved in polynomial time. As far as we know there does not exist any dataset containing different types of instances of CSFI problem, divided by complexity. At the same time there does not exist any algorithm to generate instances of CSFI that are not "easy" to solve.

In this work we present a formal approach to the definition of "trivial non-isomorphic" instances and an algorithm to generate "non-trivial" instances. This generator will allow us to both experimentally compare algorithms to solve the CSFI, and to train neural networks for the same tasks, as seen recently for some NP-complete problems such as SAT [9] and TSP [8].

2 Related Work

The problem of deciding whether two Boolean formulas are semantically isomorphic is known as the *Formula Isomorphism* (FI) problem [1,2]. Note that, from the definitions of Formula Equivalence and Formula Isomorphism problems it follows that two semantically equivalent Boolean formulas are also semantically isomorphic since the semantic equivalence relationship preserves the semantic isomorphism. Thierauf [10] showed that Graph Isomorphism (GI) is polynomial time reducible to FI, thus showing that FI is in the GI class. Later, Ausiello et al. [3] proved, as mentioned before, that CSFI belongs to GI as well.

The eponymous Graph Isomorphism problem has been studied in depth in the literature; we refer the interested reader to the classical paper of McKay [6], to the more recent work of McKay and Piperno [7], and to the quasipolynomial time algorithm of Babai [4].

3 Preliminaries and Problems Definitions

In this section we provide the necessary background and definitions of the problems considered, beginning with the CNF and monotone CNF.

Definition 1 (CNF). A formula in Conjunctive Normal Form is the conjunction of the set of clauses C, each one composed by the disjunction of a subset of the literals $L_c \subseteq L$, such that the literal could be either positive or negated. Also each literal $l \in L$ appears at most twice in each clause $c \in C$, and, if it appears twice then it should be once positive and once negated.

In this case we define the CNF formula F as composed by the set of clauses and the set of literals:

$$F = (C, L)$$

Definition 2 (Monotone CNF). A monotone Conjunctive Normal Form formula is a CNF formula as defined in Definition 1, with a restriction: all the literals in all the clauses appear always in the positive form (so there are no negated literals in formula).

We also define the problem of syntactic isomorphism as in [3]:

Definition 3 (CNF Syntactic Formula Isomorphism (CSFI)). Given two Boolean formulas in CNF as defined in Definition 1:

$$\varphi(a_1, ..., a_n) = \bigwedge_{c=1}^{m} \left(\bigvee_{l=1}^{2n} \alpha_{(c,l)} \right) = (\alpha_{(1,1)} \vee, ..., \vee \alpha_{(1,2n)}) \wedge, ..., \wedge (\alpha_{(m,1)} \vee, ..., \vee \alpha_{(m,2n)})$$

$$\varphi(b_1, ..., b_n) = \bigwedge_{c=1}^{m} \left(\bigvee_{l=1}^{2n} \beta_{(c,l)} \right) = (\beta_{(1,1)} \vee, ..., \vee \beta_{(1,2n)}) \wedge, ..., \wedge (\beta_{(m,1)} \vee, ..., \vee \beta_{(m,2n)})$$

The syntactic isomorphism problem of CNF Boolean formulas (CSFI) is to decide whether there exists a permutation of the clauses σ_c and a permutation of the literals σ_l in $\varphi(a_1, ..., a_n)$ and a bijection f between their variables such that $\varphi(a_1, ..., a_n)$ and $\varphi(b_1, ..., b_n)$ may be written in the same way i.e.:

$$\bigwedge_{c=1}^{m} \left(\bigvee_{l=1}^{2n} f(\alpha_{(\sigma_c(c), \sigma_l(l))}) \right) = \bigwedge_{c=1}^{m} \left(\bigvee_{l=1}^{2n} \beta_{(c,l)} \right)$$

As an example, considering the following formulas defined over the variables set $\{x, y, z\} \cup \{a, b, c\}$:

$$\varphi(z, y, x) = (\neg z \vee z) \wedge (y \vee x) \wedge (z \vee \neg y \vee \neg x)$$

$$\varphi(a, c, b) = (a \vee b) \wedge (\neg a \vee c \vee \neg b) \wedge (\neg c \vee c)$$

A possible solution consists in the following bijection $f = \{\langle y, a\rangle, \langle x, b\rangle, \langle z, c\rangle\}$ and a suitable permutation of σ_c and σ_l.

Definition 4 (Monotone CNF Syntactic Formula Isomorphism (MCSFI))
Considering two Boolean monotone formulas in CNF as defined in Definition 2, the syntactic isomorphism problem for CNF monotone Boolean formulas (MCSFI) is the CSFI problem restricted to CNF monotone formulas.

It's easy to note that, if we have two formulas F and F', both defined as in Definition 1, and with equal sets of both clauses and literals $F = (C, L)$, $F' = (C, L)$, then F and F' are isomorphic. In this case the isomorphism consists in a bijection $f = \{\langle l, l\rangle \mid l \in L\}$. We can define this isomorphism as trivial isomorphism.

Definition 5 (Trivial Isomorphism I). A trivial isomorphism I is an isomorphism that maps the clauses and literals sets as themselves, with also a identity bijection of the literals. So, considering $F = (C, L)$:

$$I(F) = \begin{cases} c \rightarrow c & \forall c \in C \\ l \rightarrow l & \forall l \in L \end{cases}$$

Now, to define the symmetric concept of trivial non-isomorphism we need to define some properties of the formula:

Definition 6 (Cardinality of a literal l_i in a formula F). The cardinality of a literal l_i in a formula F is the size of the set of clauses of the formula F that contain l_i:

$$|l_{i,F}| = |\{c \mid c \in F, l_i \in c\}|$$

Definition 7 (Cardinality of a clause c_i in a formula F). The cardinality of a clause c_i in a formula $F = (C, L)$ is the size of the set of literals in clause c of the formula F:

$$|c_{i,F}| = |\{l \mid l \in L, l \in c_i\}|$$

Definition 8 (Tuple of literals' cardinalities of a formula F). The tuple of literals' cardinalities of a formula F is the sorted tuple of cardinalities of all the literals in the formula F as defined in Definition 6:

$$T_F = (|l_{0,F}|, \ldots, |l_{n,F}|)$$

where $|l_{i-1,F}| \leq |l_{i,F}|$.

Definition 9 (Tuple of literals' cardinalities with respect to the formula F in clause c). The tuple of literals' cardinalities with respect to the formula F in clause c is the sorted tuple of cardinalities of the literals contained in the clause c with cardinality as defined in Definition 6:

$$T_{F,c} = (|l_{i,F}|, \ldots, |l_{j,F}|) \quad | \quad l_{i,F}, \ldots, l_{j,F} \in c$$

where $|l_{k-1,F}| \leq |l_{k,F}|$.

Definition 10 (Tuple of clauses' cardinalities of the formula F). The tuple of clauses' cardinalities with respect to the formula $F = (C, L)$ is the sorted tuple of the cardinalities of every $c \in C$ as defined in 7:

$$T_F^C = (|c_{i,F}|, \ldots, |c_{j,F}|) \quad | \quad c_{i,F}, \ldots, l_{j,F} \in C$$

where $|c_{k-1,F}| \leq |c_{k,F}|$.

Now we can define the conditions of trivial non-isomorphism:

Definition 11 (Trivial non-isomorphic). Considering two formulas $F = (C, L)$ and $F' = (C\prime, L\prime)$ as defined in Definition 1, they are trivially non-isomorphic if at least one of the following conditions is true:

$$|C| \neq |C'| \qquad |L| \neq |L'| \qquad T_F \neq T_{F'} \qquad T_F^C \neq T_{F'}^{C'}$$

Now we can define the problem we want to face:

Definition 12 (Generation of non-trivial non-isomorphic formula). Given a formula $F = (C, L)$ as defined in Definition 1, we want to find another formula $F' = (C', L')$ such that, F and F' are non isomorphic and they are not trivially non-isomorphic as defined in Definition 11.

4 The Algorithm

First of all, let's define some useful notation:

Definition 13 (Conditional clauses subsets). Given a formula $F = (C, L)$ as defined in Definition 1, we represent some subsets of the set C, based on either the presence or the absence of some literal, as:

$$\begin{aligned} C_l^F &= \{c \mid c \in C,\, l \in c\} \\ C_{l_1,l_2}^F &= \{c \mid c \in C,\, l_1 \in c,\, l_2 \in c,\, l_1, l_2 \in L\} \\ C_{l_1,\not{l_2}}^F &= \{c \mid c \in C,\, l_1 \in c,\, l_2 \notin c,\, l_1, l_2 \in L\} \end{aligned}$$

We also define a function to map a formula (or a sub-formula) to a vector space based on the literals' cardinalities:

Definition 14 (Cardinalities vector space). Given a formula $F = (C, L)$ as defined in Definition 1, we define a function v that maps F to a vector space based on the cardinalities of the literals contained in L. The function gives in output a set (with repetitions) of vectors, composed by cardinalities.

$$\begin{aligned} v(F, C') &= \{T_{F,c} \mid c \in C'\} \\ v(F, C', \not{l}) &= \{T_{F',c} \mid c \in C' \;,\; F' = (C, L - \{l\})\} \end{aligned}$$

with $C' \subseteq C$ and $l \in L$.

To apply the algorithm, that will be defined later in this section, we need to find two literals to apply the algorithm to. So we define the type of literals that we need:

Definition 15 (Asymmetrical literals). Given a formula $F = (C, L)$ as defined in Definition 1, we define a two literals $\alpha,\ \beta \in L$ as asymmetrical literals if they respect all the following conditions:

$$|\alpha_F| > |\beta_F| \qquad |C_{\alpha,\not{\beta}}^F| > 0 \qquad |C_{\beta,\not{\alpha}}^F| > 0$$
$$\exists u \mid T_{F,u} \in v(F, C_{\alpha,\not{\beta}}^F, \not{\alpha}),\ T_{F,u} \notin v(F, C_{\beta,\not{\alpha}}^F, \not{\beta})$$

Now we can define the algorithm:

Algorithm 1 (Monotone non-isomorphic CNF generator). Given a monotone CNF formula $F = (C, L)$ as defined in definition 2 that contains two literals $\alpha, \beta \in L$ such that α and β are asymmetrical literals as defined in 15. Now, given that:

- $\delta = |C^F_\alpha| - |C^F_\beta|$
- u is the clause as in the definition 15 of the asymmetrical literals
- D is any set such that $D \subseteq C^F_{\alpha,\not\beta} - \{u\}, \quad |D| = \delta$

then the new formula F' will be composed as $F' = (C', L)$ such that $C' \equiv (C \setminus D) \cup D'$, where D' is the same set of clauses contained in D except that all the occurrences of α are replaced with β.

Now we can analyse the result of the algorithm:

Theorem 1. *Given a monotone CNF formula F as defined in Definition 2 and F' as the result of Algorithm 1 applied over F, then F and F' are not isomorphic and they are not trivially non-isomorphic as defined in Definition 11.*

Proof. The resulting formula $F' = (C', L)$ will be equal to F for every clause c such that $c \notin C'_{\alpha,\not\beta}$, $c \notin C'_{\beta,\not\alpha}$. Since that we had just transformed some instances of α to β, it means that we have changed only $C_{\alpha,\not\beta}$ and $C_{\beta,\not\alpha}$. And we changed only δ instances, so we can see that

$$|C'_{\alpha,\not\beta}| = |C_{\alpha,\not\beta}| - \delta \qquad |C'_{\beta,\not\alpha}| = |C_{\beta,\not\alpha}| + \delta$$

but we also know that

$$|C_{\alpha,\not\beta}| = |C_{\beta,\not\alpha}| + \delta$$

so, it means that

$$|C'_{\alpha,\not\beta}| = |C_{\beta,\not\alpha}| + \delta - \delta \qquad |C'_{\beta,\not\alpha}| = |C_{\alpha,\not\beta}| - \delta + \delta$$

and so:

$$|C'_{\alpha,\not\beta}| = |C_{\beta,\not\alpha}| \qquad |C'_{\beta,\not\alpha}| = |C_{\alpha,\not\beta}|$$

From the previous equations it follows that all the following are true:

$$|C| = |C'| \qquad |L| = |L'| \qquad T_F = T_{F'} \qquad T^C_F = T^{C'}_{F'}$$

so, from the Definition 11 we know that F and F' are not trivially non-isomorphic. Now we have to prove that F and F' are anyway non-isomorphic. We know that a valid mapping should map each literal to a literal with the same cardinality in the formula, so it also means that each clause should be mapped to a clause with an equal sorted tuple of cardinalities. And since it should be true for each clause, it means that the two formulas should have the same set (with repetitions) of sorted tuples of cardinalities. And so it means that to have two formulas $F = (C, L)$ and $F' = (C', L')$ isomorphic, it is necessary to have $v(F, C) \equiv v(F', C')$

Now, let's consider the mapping that we have defined before. In this mapping we changed only δ clauses, but we also swapped the cardinalities of two literals, so we could have affected the result of the function v for all the clauses that contain either α or β. So let's try to analyse the four sets separately:

A $C_{\alpha,\beta}$
B $C'_{\beta,\not\alpha} - C_{\beta,\not\alpha}$: the set containing only the δ clauses changed
C $C'_{\beta,\not\alpha} \cap C_{\beta,\not\alpha}$: the set containing only the β clauses that didn't change
D $C'_{\alpha,\not\beta} \cap C_{\alpha,\not\beta}$: the set containing only the α clauses that didn't change

Set A: In the set $C_{\alpha,\beta}$ we just swapped the cardinalities, so the results will be the same for the function v.
Set B: In this clauses we changed α with β, but at the same time $|\alpha_F|$ is equal to $|\beta_{F'}|$. So the results will be the same for the function v.
Sets C and D: Since that the clauses in these sets were not changed, they will be equal, but the cardinalities of α and β will be changed. Since that we have chosen clauses that are different from s, it means that s is still the same, but it cardinality is changed, so:

$$v(F, \{u\}) \neq v(F', \{u\})$$

but by definition $v(F, \{u\}, \not\alpha)$ is in $C_{\alpha,\not\beta}$ but it is not in $C_{\beta,\not\alpha}$, so it means that it is not possible to have some other clause u' such that

$$v(F, \{u\}) = v(F', \{u'\})$$

The result is that $v(F) \neq v(F')$ and so F and F' are non-isomorphic. □
To extend the result to any CNF we need to define some other notation:

Definition 16 (Tuple of cardinalities of a literal l with respect to the formula F). Tuple of cardinilities of literal with respect to the formula F for each literal $l \in L$ of a formula $F = (C, L)$ as defined in Definition 1, is the tuple composed by the sorted cardinalities of both positive and negated literal l:

$$|l_F|^* = (|l_F|_m, |l_F|_M) = \begin{cases} (|l_F|, |\neg l_F|) & \text{if } |l_F| \leq |\neg l_F| \\ (|\neg l_F|, |l_F|) & \text{otherwise} \end{cases}$$

Definition 17 (Tuple of positive and negated cardinalities of a formula F). Given a formula $F = (C, L)$ as defined in Definition 1, the tuple of positive and negated cardinalities of a formula F is the tuple of all the tuples of cardinalities of the literals in L:

$$TN_F = \{|l_F|^* \mid l \in L\}$$

such that it is always valid that

$$\begin{cases} |l_{i,F}|_m < |l_{j,F}|_m & \text{if } |l_{i,F}|_M = |l_{j,F}|_M \\ |l_{i,F}|_M < |l_{j,F}|_M & \text{otherwise} \end{cases}$$

for any $i < j$

Now we can also extend the definition of trivial non-isomorphism:

Definition 18 (Trivial non-isomorphic for general cnf). Considering two formulas $F = (C, L)$ and $F' = (C, L)$ as defined in Definition 1, they are trivially non-isomorphic if at least one of the following conditions it's true:

$$|C| \neq |C'| \qquad |L| \neq |L'| \qquad TN_F \neq TN_{F'}$$

If we consider two literals α and β in a formula $F = (C, L)$, we can have only one of the following cases:

1. $|\alpha_F| = |\beta_F|,\ |\neg\alpha_F| = |\neg\beta_F|$
2. $|\alpha_F| \neq |\beta_F|,\ |\neg\alpha_F| = |\neg\beta_F|$
3. $|\alpha_F| = |\beta_F|,\ |\neg\alpha_F| \neq |\neg\beta_F|$
4. $|\alpha_F| \neq |\beta_F|,\ |\neg\alpha_F| \neq |\neg\beta_F|,\ |\neg\alpha_F| \neq |\beta_F|,\ |\alpha_F| = |\neg\beta_F|$
5. $|\alpha_F| \neq |\beta_F|,\ |\neg\alpha_F| \neq |\neg\beta_F|,\ |\neg\alpha_F| = |\beta_F|,\ |\alpha_F| \neq |\neg\beta_F|$
6. $|\alpha_F| \neq |\beta_F|,\ |\neg\alpha_F| \neq |\neg\beta_F|,\ |\neg\alpha_F| \neq |\beta_F|,\ |\alpha_F| \neq |\neg\beta_F|$

Since that we could apply the following mappings:

- $\alpha \to \neg\alpha',\ \beta \to \neg\beta'$ to transform the case 3 in the case 2
- $\alpha \to \neg\alpha'$ to transform the case 4 in the case 2
- $\beta \to \neg\beta'$ to transform the case 5 in the case 2

then we can consider, without losing of generality, to have only the following cases:

1. $|\alpha_F| = |\beta_F|,\ |\neg\alpha_F| = |\neg\beta_F|$
2. $|\alpha_F| \neq |\beta_F|,\ |\neg\alpha_F| = |\neg\beta_F|$
3. $|\alpha_F| \neq |\beta_F|,\ |\neg\alpha_F| \neq |\neg\beta_F|,\ |\neg\alpha_F| \neq |\beta_F|,\ |\alpha_F| \neq |\neg\beta_F|$

Now we can define a more general version of asymmetrical literals:

Definition 19 (Asymmetrical literals of positive or negated literals). Given a formula $F = (C, L)$ as defined in Definition 1, we define a two literals $\alpha,\ \beta \in L$ as asymmetrical literals if they respect all the following conditions:

$$|\alpha_F| > |\beta_F| \qquad |C^F_{\alpha,\not\beta}| > 0 \qquad |C^F_{\beta,\not\alpha}| > 0 \qquad |\neg\alpha_F| = |\neg\beta_F|$$
$$\exists u \mid T_{F,u} \in v(F, C^F_{\alpha,\not\beta}, \not\alpha),\ T_{F,u} \notin v(F, C^F_{\beta,\not\alpha}, \not\beta)$$

Now we have all the information to define a more general algorithm:

Algorithm 2 (Non-isomorphic cnf generator). Given a monotone cnf formula $F = (C, L)$ as defined in definition 1, we take two asymmetrical literals $\alpha, \beta \in L$ as defined in 19. Now, given that:

- $\delta = |C^F_\alpha| - |C^F_\beta|$
- u is the clause as in the definition 19 of the asymmetrical literals
- D is any set such that $D \subseteq C^F_{\alpha,\not\beta} - \{u\},\quad |D| = \delta$

then the new formula F' will be composed as $F' = (C', L)$ such that

$$C' \equiv C \setminus D \cup D'$$

where D' is the same set of clauses contained in D except that all the occurrences of α are replaced with β.

So we can analyse the result of the algorithm:

Theorem 2. *Given a cnf formula F as defined in Definition 1 and F' as the result of the Algorithm 2 applied over F, then F and F' are not isomorphic and they are not trivially non-isomorphic as defined in Definition 18.*

Proof. It's easy to verify that the negative literals are not involved in the mapping and they remain as they were before. It means that their cardinalities will be swapped but by definition $|\neg\alpha_F| = |\neg\beta_F|$, and they will not change with respect to the previous formula. So we could replace all the negated literals with some other new literal to have all the literals independent to each other. So we are in the same exact case of the Theorem 1. □

Since that the differences between the instance F' built from Algorithm 2 and the original CNF F in input, could be very small, it could be useful to apply a simple isomorphism to F' to generate a new formula F''. Since that F and F' are not isomorphic then also F and F'' are not isomorphic, and they also are non-trivially non-isomorphic, keeping at the same time obfuscated the small differences between F and F'.

5 Conclusion

In this paper we have shown that there are different types of syntactic non-isomorphism with respect to the complexity of the isomorphism testing. Then we have shown an algorithm that, given a CNF, generates a new CNF that is non-trivially non-isomorphic to the original CNF. The implementation of this algorithm, together with other CNF utilities, can be found on https://github.com/paolofantozzi/cnf-generator.

The results showed in this work are needed to build a dataset that will be used to estimate the complexity of testing an instance of a problem. In this case we can distinguish between trivially non-isomorph istances and non-trivially non-isomorph instances.

Our goal is to be able to use the generator to train neural networks able to solve CSFI, such in the recent works, for NP-complete problems, of Selsam et al. [9], that trained a neural network that learns to solve SAT problems after only being trained as a classifier to predict satisfiability, and Prates et al. [8], that showed that Graph Neural Networks can learn to solve, with very little supervision, the decision variant of the Traveling Salesperson Problem (TSP). Some preliminary results of training a neural network model, using the generator presented in this work, are shown in [5].

References

1. Agrawal, M., Thierauf, T.: The Boolean isomorphism problem. In: Proceedings of 37th Conference on Foundations of Computer Science (FOCS), pp. 422–430. IEEE (1996)
2. Agrawal, M., Thierauf, T.: The formula isomorphism problem. SIAM J. Comput. **30**(3), 990–1009 (2000)
3. Ausiello, G., Cristiano, F., Fantozzi, P., Laura, L.: Syntactic isomorphism of CNF Boolean formulas is graph isomorphism complete. In: Cordasco, G., Gargano, L., Rescigno, A.A. (eds.), Proceedings of the 21st Italian Conference on Theoretical Computer Science, Ischia, Italy, 14–16 September 2020, CEUR Workshop Proceedings, vol. 2756, pp. 190–201 (2020). CEUR-WS.org
4. Babai, L.: Graph isomorphism in quasipolynomial time [extended abstract]. In: Proceedings of the 48th Annual ACM SIGACT Symposium on Theory of Computing - STOC 2016. ACM Press, New York (2016)
5. Benedetto, L., Fantozzi, P., Laura, L.: Complexity-based partitioning of CSFI problem instances with transformers (2021)
6. McKay, B.D., et al.: Practical graph isomorphism (1981)
7. McKay, B.D., Piperno, A.: Practical graph isomorphism. II. J. Symbol. Comput. **60**, 94–112 (2014)
8. Prates, M., Avelar, P.H.C., Lemos, H., Lamb, L.C., Vardi, M.Y.: Learning to solve NP-complete problems: a graph neural network for decision TSP. In: AAAI, vol. 33, no. 01, pp. 4731–4738 (2019)
9. Selsam, D., Lamm, M., Bünz, B., Liang, P., de Moura, L., Dill, D.L.: Learning a SAT solver from single-bit supervision, February 2018
10. Thierauf, T.: The Computational Complexity of Equivalence and Isomorphism Problems. Springer, Heidelberg (2000). https://doi.org/10.1007/3-540-45303-2

Check for updates

A Search Engine for Scientific Publications: A Cybersecurity Case Study

Nuno Oliveira(✉), Norberto Sousa, and Isabel Praça

Research Group on Intelligent Engineering and Computing for Advanced Innovation and Development (GECAD), Porto School of Engineering (ISEP), 4200-072 Porto, Portugal
{nunal,norbe,icp}@isep.ipp.pt

Abstract. Cybersecurity is a very challenging topic of research nowadays, as digitalization increases the interaction of people, software and services on the Internet by means of technology devices and networks connected to it. The field is broad and has a lot of unexplored ground under numerous disciplines such as management, psychology, and data science. Its large disciplinary spectrum and many significant research topics generate a considerable amount of information, making it hard for us to find what we are looking for when researching a particular subject. This work proposes a new search engine for scientific publications which combines both information retrieval and reading comprehension algorithms to extract answers from a collection of domain-specific documents. The proposed solution although being applied to the context of cybersecurity exhibited great generalization capabilities and can be easily adapted to perform under other distinct knowledge domains.

1 Introduction

Cybersecurity is a neoteric field that emerged out of the latest advances in computer science [1]. Although there is not yet a consensual agreement between the scientific community across the whole scope of cybersecurity research topics, some works have tried to systematize research categories [1,2], being one of them related to data science applications.

The recent developments in software, hardware, and network topologies contributed to more complex systems such as Cyber-Physical Systems (CPS) in which the capabilities of computing, communications, and data storage are used to monitor physical and cyber entities [3]. Furthermore, these advances can also be translated into more sophisticated cyberattacks comprised of multiple attack vectors. Hence, the complex nature of cyber threats and the need to progressively adapt security systems to the most relevant ones makes the application of Artificial Intelligence (AI) a promising technology to use for increased cybersecurity [4].

Being cybersecurity such a hot research topic nowadays, with so many different applications, it is hard to efficiently find answers to specific topics in the

© The Author(s), under exclusive license to Springer Nature Switzerland AG 2022
K. Matsui et al. (Eds.): DCAI 2021, LNNS 327, pp. 108–118, 2022.
https://doi.org/10.1007/978-3-030-86261-9_11

wide amount of existing scientific publications. Natural Language Processing (NLP) methods, namely Reading Comprehension (RC) algorithms, can give a substantial contribution to solving the introduced problem. However, the ability to read a text and then answer questions about it is a very difficult task for machines [5].

Over the last few years, the introduction of reliable data collections such as the Stanford question answering dataset (SQuAD) [5] and the development of deep learning methods based on transformer architectures [6] such as BERT [7] and RoBERTa [8] have contributed to major improvements in the field of RC. Nevertheless, it is not feasible to apply these algorithms directly to huge amounts of text due to computational limits and performance issues. To overcome this problem, Information Retrieval (IR) methods [9] can be used to measure the relevance of a given document to a given question providing a filter to find only relevant data and narrowing down the search space.

This work proposes a novel Question Answering (Q&A) system for the cybersecurity research context in which one can place a domain-related question and expect a direct answer retrieved from a set of scientific publications. This system uses a combination of IR and RC methods to perform the task described above and provides the results in a user-friendly web interface. Although the fact that the cybersecurity context is considered for the case study, the proposed method can be easily applied to many other different domains.

This work is organized in multiple sections that can be detailed as follows. Section 2 provides an overview of current information retrieval and reading comprehension algorithms and applications. Section 3 describes the proposed solution, detailing both the software architecture and employed algorithms. In Sect. 4, our solution is applied to the case study and the obtained results are presented and discussed. Section 5 provides a summary of the main conclusions that can be drawn from this research and appoints further research topics to be addressed in the future.

2 Related Work

Due to the multiple domains intelligent QA systems are connected to, we will analyze the literature on multiple different subjects. One of such subjects is text mining and document ranking systems, of which the internet and search engines are a great example [10]. Taking into account the scope of our work, we investigated weighing methods such as Term Frequency - Inverse Document Frequency (TF-IDF), Dense Passage Retrievers (DPR), and word embeddings.

In [11], Shahzad Qaiser *et al.*, employs a TF-IDF ranking system to several web pages in order to compare results. TF-IDF is the most utilized weighting scheme for web searches of information retrieval and text mining [12]. The author also points TF-IDF's biggest issue, which is not identifying different tenses of words. In the same manner, Joel L. Neto *et al.* in [13] employs a modified version of TF-IDF, TF-ISF, applying stemming to reduce the impact of this classification method's weaknesses.

In [14], Karpukhin and Oğus *et al.*, utilized the standard BERT pre-trained model and a DPR in a dual encoder architecture achieving state of the art results. Their DPR exceeds BM25's capabilities by far, namely a more than 20% increase in top-5 accuracy (65.2%). Their results for end-to-end QA accuracy also improved on ORQA, the first open-retrieval question answering system, introduced in [15] by Lee *et al.*, in the natural questions dataset [16].

Regarding word embedding, in which a document's words are mapped as vectors in a continuous vector space, words with similar meanings will be closer to one another, aiding in dimensionality reduction [17]. In [18], Tomas Mikolov *et al.* demonstrates the application of a skip-gram model, a more computational efficient architecture, to mapping words to a vectorial space, and the same model but focusing on phrases.

On the other hand, the Q&A task involves the search for relationships and meaning between entities. Due to the nature of language, this search becomes extremely complex, given that context can change the meaning of any sequence of words. In NLP, the key to solve entity-related tasks is to create a model to learn the optimal way of entity representation.

Ordinarily, each entity in the Knowledge Base (KB) is assigned an embedding vector, capturing information in it. Due to the scope restriction of this method, entities that are outside of the KB are not represented and therefore any model built on top of it performs poorly.

To solve this issue, Contextual Word Representations (CWR) are employed with generalized word representations that serve multiple purposes. These CWRs are based on the transformer architecture, most notably BERT [7] and following improvements such as RoBERTa [8] that perform extremely well in a wide range of NLP tasks such as document classification and entanglement, sentiment analysis, question answering, sentence similarity, etc. These representations are obtained by training a model on a large-scale corpus (ex: Wikipedia) and can then be transferred to other network-based models, allowing them to improve search-related tasks such as the relevant question context as shown by Wei Yang *et al.* in [19] where, in an end-to-end QA system, the integration of BERT outperformed previous implementations by significant margins.

3 Proposed Solution

To solve the introduced problem we built a prototype using the python programming language on top of the haystack framework [20]. The system was designed as a client-server architecture with two main components, the front-end, a web-based graphical interface that can be accessed by the users and the back-end, a RESTful API that exposes the use cases of our solution through several endpoints. Additionally, there is also an SQLite database which is used to store preprocessed scientific articles.

The back-end side of our application can also be further detailed into two distinct modules, a web-crawler, which is integrated with arXiv.org API so that it can fetch scientific articles in real time, and a search engine, which combines two distinct NLP methods, a retriever and a reader, to build a pipeline that is able to find candidate answers in our corpus to user-specified questions. The described architecture is represented in Fig. 1.

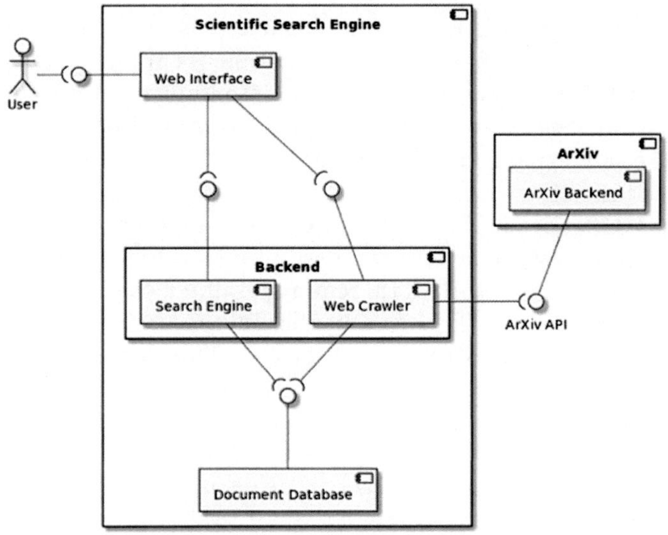

Fig. 1. Solution architecture.

The proposed system regards three main use cases that can be described as follows:

- **UC1 - Download Scientific Articles**: The user specifies a given search topic and the maximum number of articles to be downloaded. The crawler tries to find articles related to the specified topic and downloads all articles until the maximum limit is reached. Then, the documents are preprocessed - empty lines are removed, consecutive whitespaces are truncated and pdf headers and footers are discarded. After preprocessing, the text of each document is split into several search chunks of 500 words (respecting sentence boundary) so that the search process can be optimal. Finally, each resulting chunk is indexed, along with the document meta-data, in the document database.
- **UC2 - Consult Database Summary**: In the main dashboard of the graphical interface a summary of the document database content is displayed so that the user can keep track of the changes to the available corpus. The database summary is comprised of several data points such as the number of downloaded articles, search chunks and document categories.
- **UC3 - Find Candidate Answers**: The user places a question to the system and specifies several search parameters such as a category filter, the number of candidate answers to be displayed, *c*, and the maximum number of relevant

search chunks to be found by the retriever, k. The system will first execute the retriever, a TF-IDF-based retriever which will return the most relevant k chunks. Then, the reader, a RoBERTa model, will try to find the best c answers in the selected k chunks according to a confidence metric.

The described solution is quite generic since it is easy to enrich the search corpus with the contents of scientific publications of different subjects due to the execution of **UC1**. However, for this concrete implementation it would only be possible to find articles stored in the arXiv.org repository. Nonetheless, this feature can be easily expanded by integrating the existing web crawler with other scientific repositories.

Regarding **UC3**, the proposed NLP pipeline is also quite broad, the retriever, TF-IDF is not context-specific and can easily be used for multiple domains. On the other hand, the reader, RoBERTa, requires training examples comprising different questions and answers. To overcome this limitation, we opted to use a model that was pre-trained on the SQuAD dataset [21]. This data collection comprises over 100,000 examples of questions posed by crowdworkers on a set of Wikipedia articles [5] resembling a good benchmark dataset for training and evaluating general-purpose extractive Q&A machine learning models. The RoBERTa model employed in our solution, [21], achieved an exact match score of approximately 79.97% and an f1-score of 83.00% under this testbed. In our experiments, the search engine performed quite competently being able to find interesting answers to several questions that were placed regarding the cybersecurity domain.

It is possible to further improve the proposed solution by adding new functionalities regarding the database management, namely, to perform listings of downloaded articles accordingly to a combination of search criteria, to manually import a given scientific article and to delete unwanted articles.

3.1 Pipeline Description

All steps of the search pipeline are described in this section.

3.1.1 Retriever

In order to search through relevant information, a TF-IDF retriever was put in place. It is a numerical statistic that is intended to reflect how important a given word is to a document in a corpus.

$$TFIDF_{i,d} = tf_{i,d} \cdot idf_i \tag{1}$$

In the scientific question and answering domain, it is expected that the queries will have lexical overlap with their answers, making this algorithm a good searcher of relevant information.

3.1.2 Reader

Another critical step of our pipeline is the question understanding step. Here we need to be able to properly understand the question at hand. By being able to properly model it in such a way that it can then be passed through the pipeline and improving the chances of getting not only accurate but also relevant answers. For this step, we use a FARM reader coupled with the RoBERTa [8] language model which works alongside the retriever and parses the candidate documents provided. RoBERTa is an iteration of the BERT [7] language model whose architecture is based on the Transformer architecture, Fig. 2. This new architecture disregards recurrence and convolutions from the usual encoder-decoder models and instead focuses on several types of attention mechanisms. It introduces several novelties such as sclaled-dot product attention, multi-head attention and positional encoding. At each time step the output of the decoder stack is fed back to the decoder similarly to how the outputs of previous time steps are used as hidden states in Recurrent Neural Networks (RNN) [6].

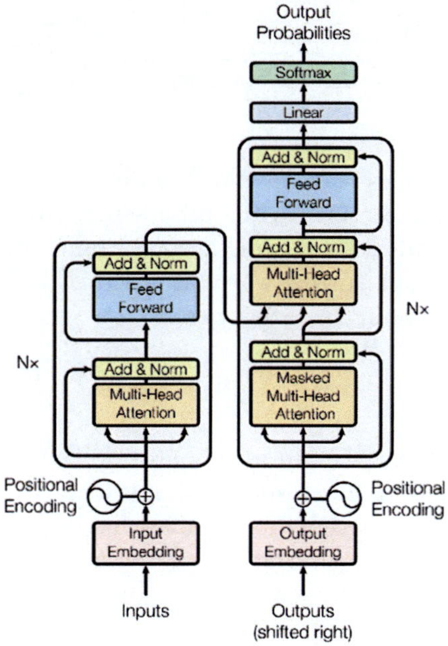

Fig. 2. The transformer architecture [6].

RoBERTa is also trained on a much larger corpus than BERT and as a result, achieves significant performance gains.

4 Case Study

Despite the usefulness and generalization of our solution, which allows it to be applied to numerous topics, for our case study we have decided to focus on a current and challenging research topic - cybersecurity. For this reason we compiled a list of keywords related to that topic that we used to find relevant articles to build our search corpus. For each keyword we obtained a number of articles as shown in Table 1. After removing the corrupted/unparsable documents and duplicates, our corpus totalled 821 articles.

Table 1. Corpus composition.

Adversarial Attack	200
Attack Detection	175
Cyberphysical Systems	200
Cybersecurity	129
Intrusion Detection Systems	130
Total Used	834
Corrupted	−6
Duplicates	−7
Total Articles in Corpus	821

Each one of these articles was downloaded and processed as per the pipeline indicated in the previous section. After processing, the articles were split into chunks of 500 words while taking into account sentence continuity. With the finalization of this step, our corpus was composed of 12827 search chunks from 821 different articles of about 36 categories.

4.1 Results

The introduced solution has a main dashboard, on the left some search configuration sliders and database related information is located. In the middle there are two buttons to navigate between the database management and search engine functionalities. The described interface is presented in Fig. 3.

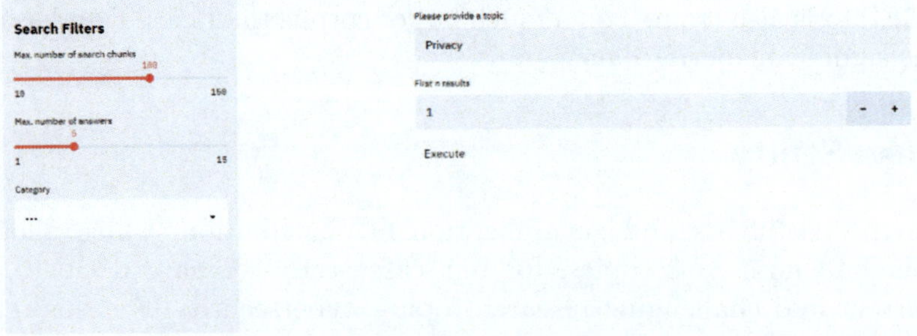

Fig. 3. Dashboard.

With the corpus prepared, it is then possible to start asking questions. By asking: "What are the challenges of AI?", the most interesting candidate answer is presented in Fig. 4, due to its high probability (confidence) score. This answer is highlighted in its surrounding context, accompanied by additional information such as title, authors, publishing date, and a link to the article itself.

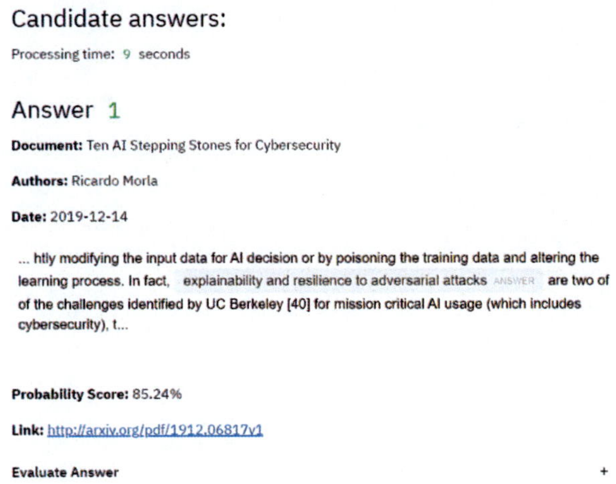

Fig. 4. Question example.

As the question is vague in nature, and the prepared corpus is geared more towards cybersecurity instead of AI, the obtained answer "explainability and resilience to adversarial attacks" also tends to the cybersecurity side of AI, due to the nature of the used article [22].

Another example is the question, "What are the main challenges of cybersecurity research?" which yielded interesting results. The first answer correctly quotes [23] and responds with "lack of adequate evaluation/test environments that utilize up-to-date datasets, variety of testbeds while adapting unified evaluation methods", while the second answer builds on the first one with "lack of research methodology standards" [24].

Finally, by asking "Which machine learning models are commonly used?" we obtain "Naïve Bayes, SVM, KNN, and decision trees" from [25] and virtually the same answer "Support Vector Machine, Decision Trees, Fuzzy Logic, BayesNet and Naïve Bayes" from [26].

The quality of the responses found is directly connected to the contents of the corpus. This can be remedied by populating the corpus with more articles pertaining to a given topic or adding a new topic entirely. For this we can access the database management functionality, and specify a given search topic and the maximum number of documents to be downloaded. These will be directly fetched from arXiv.org, preprocessed and indexed alongside their metadata in the document database.

For the topic of “Privacy”, with a maximum of one article, the result is presented in Fig. 5.

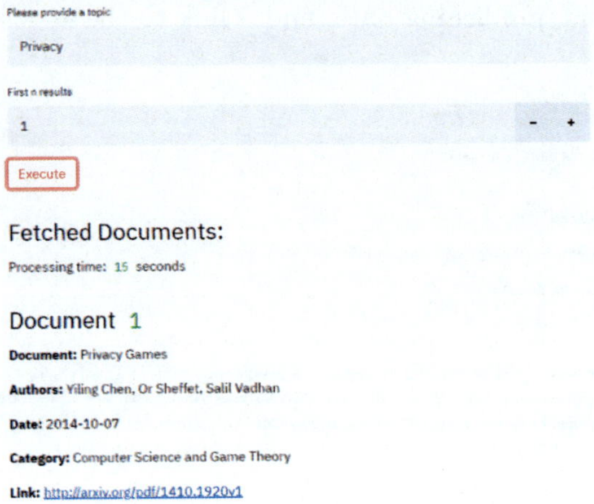

Fig. 5. Database menu.

Our solution for the cybersecurity use case performed admirably, by compiling a corpus of 821 articles on five of the hottest research topics in the field and by finding interesting answers to a set of significant questions regarding applications of AI to cybersecurity and the main challenges of current research. Regarding the extractive Q&A pipeline, the RoBERTa model exhibited a notable adaptation capability since it was not retrained in the scope of the cybersecurity scientific domain.

5 Conclusion

Given the amount of scientific articles that are published every year it is hard to find exactly what we are looking for when researching a particular topic. In this work, we have presented a software solution that aims to solve this problem. It comprises several advantageous features such as the continuous update of the search corpus by providing an easy-to-use integration with the arXiv.org API and the ability to find candidate answers extracted from the corpora of downloaded scientific publications by applying a combination of two NLP methods, TF-IDF and RoBERTa.

Furthermore, the introduced solution was showcased in the context of cybersecurity, a neoteric field of science with increasing interest. With a base corpus of 821 articles, the system was able to find proper answers to questions such as “What are the challenges of AI?”, “What are the main challenges of cybersecurity research?” and “Which machine learning models are commonly used””
showing a great capability of generalization.

As future work, we will implement additional features regarding the document database management, expand the web crawler so that it can work with more scientific repositories and improve the document preprocessing step to make our search engine more efficient.

Acknowledgements. The present work has been developed under the EUREKA ITEA3 Project CyberFactory#1 (ITEA-17032) and Project CyberFactory#1PT (ANI|P2020 40124) co-funded by Portugal 2020.

References

1. Suryotrisongko, H., Musashi, Y.: Review of cybersecurity research topics, taxonomy and challenges: Interdisciplinary perspective. In: 2019 IEEE 12th Conference on Service-Oriented Computing and Applications (SOCA), pp. 162–167 (2019)
2. Lu, Y.: Cybersecurity research: a review of current research topics. J. Ind. Integration Manag. **03**, 08 (2018)
3. Rawung, R.H., Putrada, A.G.: Cyber physical system: paper survey. In: 2014 International Conference on ICT For Smart Society (ICISS), pp. 273–278 (2014)
4. Wirkuttis, N., Klein, H.: Artificial intelligence in cybersecurity. Cyber Intell. Secur. J. **1**(1), 21–23 (2017)
5. Rajpurkar, P., Zhang, J., Lopyrev, K., Liang, P.: SQuAD: 100,000+ questions for machine comprehension of text. In: Proceedings of the 2016 Conference on Empirical Methods in Natural Language Processing, (Austin, Texas), pp. 2383–2392. Association for Computational Linguistics, November 2016
6. Vaswani, A., et al.: Attention is all you need. In: Proceedings of the 31st International Conference on Neural Information Processing Systems, NIPS 2017, (Red Hook, NY, USA), pp. 6000–6010. Curran Associates Inc. (2017)
7. Devlin, J., Chang, M.-W., Lee, K., Toutanova, K.: BERT: pre-training of deep bidirectional transformers for language understanding. In: Proceedings of the 2019 Conference of the North American Chapter of the Association for Computational Linguistics: Human Language Technologies, Volume 1 (Long and Short Papers), (Minneapolis, Minnesota), pp. 4171–4186. Association for Computational Linguistics, June 2019
8. Liu, Y.: RoBERTa: a robustly optimized BERT pretraining approach. arXiv:1907.11692 (2019)
9. Aggarwal, C.C., Zhai, C.: A survey of text classification algorithms. In: Aggarwal, C., Zhai, C. (eds.) Mining Text Data. Springer, Boston (2012). https://doi.org/10.1007/978-1-4614-3223-4_6
10. Singh, A.K., Kumar, P.R.: A comparative study of page ranking algorithms for information retrieval. Int. J. Electr. Comput. Eng. **4**, 469–480 (2009)
11. Qaiser, S., Ali, R.: Text mining: use of TF-IDF to examine the relevance of words to documents. Int. J. Comput. Appl. **181**(1), 25–29 (2018)
12. Beel, J., Gipp, B., Langer, S., Breitinger, C.: Research-paper recommender systems?: a literature survey. Int. J. Digit. Libr. **17**(4), 305–338 (2016)
13. Neto, J.A., Santos, A.D., Kaestner, C.A., Freitas, A.A.: Document clustering and text summarization. In: Proceedings of the Fourth International Conference on the Practical Application of Knowledge Discovery and Data Mining, pp. 41–55. The Practical Application Company (2000)

14. Karpukhin, V., et al.: Dense passage retrieval for open-domain question answering. arXiv preprint arXiv:2004.04906 (2020)
15. Lee, K., Chang, M.-W., Toutanova, K.: Latent retrieval for weakly supervised open domain question answering. arXiv preprint arXiv:1906.00300 (2019)
16. Kwiatkowski, T., et al.: Natural questions: a benchmark for question answering research. Trans. Assoc. Comput. Linguist. **7**, 453–466 (2019)
17. Ge, L., Moh, T.: Improving text classification with word embedding. In: 2017 IEEE International Conference on Big Data (Big Data), pp. 1796–1805 (2017)
18. Mikolov, T., Sutskever, I., Chen, J., Corrado, G., Dean, J.: Distributed representations of words and phrases and their compositionality. arXiv preprint arXiv:1310.4546 (2013)
19. Yang, W., et al.: End-to-end open-domain question answering with bertserini. arXiv preprint arXiv:1902.01718 (2019)
20. Haystack (2020). https://haystack.deepset.ai/. Accessed 06 June 2021
21. Branden Chan, M.P., Möller, T., Soni, T.: Deepset roberta-base-squad2. https://huggingface.co/deepset/roberta-base-squad2. Accessed 06 May 2021
22. Morla, R.: Ten AI stepping stones for cybersecurity. arXiv:1912.06817 (2019)
23. Kayan, H., Nunes, M., Rana, O., Burnap, P., Perera, C.: Cybersecurity of industrial cyber-physical systems: a review, January 2021. arXiv e-prints arXiv:2101.03564
24. Gardner, C., Waliga, A., Thaw, D., Churchman, S.: Using camouflaged cyber simulations as a model to ensure validity in cybersecurity experimentation. arXiv:1905.07059 (2019)
25. Priya, V., Thaseen, I.S., Gadekallu, T.R., Aboudaif, M.K., Nasr, E.A.: Robust attack detection approach for IIoT using ensemble classifier. Comput. Mater. Continua **66**(3), 2457–2470 (2021)
26. Shah, S.A.R., Issac, B.: Performance comparison of intrusion detection systems and application of machine learning to SNORT system. Future Gener. Comput. Syst. **80**, 157–170 (2018)

Prediction Models for Coronary Heart Disease

Cristiana Neto[1], Diana Ferreira[1], José Ramos[2], Sandro Cruz[2], Joaquim Oliveira[2], António Abelha[1], and José Machado[1](✉)

[1] Algoritmi Research Center, University of Minho, Campus Gualtar, 4704-553 Braga, Portugal
{cristiana.neto,diana.ferreira}@algoritmi.uminho.pt,
{abelha,jmac}@di.uminho.pt

[2] University of Minho, 4704-553 Braga, Portugal
{a73855,pg41906,pg38931}@alunos.uminho.pt

Abstract. In the current days, it is known that a great amount of effort is being applied to improving healthcare with the use of Artificial Intelligence technologies in order to assist healthcare professionals in the decision-making process. One of the most important field in healthcare diagnoses is the identification of Coronary Heart Disease since it has a high mortality rate worldwide. This disease occurs when the heart's arteries are incapable of providing enough oxygen-rich blood to the heart. Thus, this study attempts to develop Data Mining models, using Machine Learning algorithms, capable of predicting, based on patients' data, if a patient is at risk of developing any kind of Coronary Heart Disease within the next 10 years. To achieve this goal, the study was conducted by the CRISP-DM methodology and using the RapidMiner software. The best model was obtained using the Decision Tree algorithm and with Cross-Validation as the sampling method, obtaining an accuracy of 0.884, an AUC value of 0.942 and an F1-Score of 0.881.

1 Introduction

According to the American Center for Disease Control and Prevention (CDC), heart diseases are one of the major causes of death in the world [6]. This kind of condition presents itself in a variety of forms, each posing as a major health risk ultimately leading to disability and death. There is a great array of variables that may elevate the risk of someone developing a heart condition. It is known that some personal behaviors have negative effects on human's health, catalyzing the appearance of such problems. Every day lots of people are diagnosed with heart diseases. According to the severity of the condition, after the diagnosis, the medical guideline varies from the adjustment of the patient's dietary habits to, in worse situations, having to perform a heart surgery [1]. In either case, the future of these patients is affected by having to take precautions according to their heart condition. In severe cases of the disease or the misguiding of medical recommendations can lead to the patients' death. However, if the prediction of heart conditions based on the current life-style and health conditions of the patient can be performed, it becomes possible to act preemptively, reducing the risk of developing this kind of problem. An adaptation of the lifestyle over a sufficiently long time can help to prevent heart diseases from appearing [15]. Focusing on the problem at hand, this study

© The Author(s), under exclusive license to Springer Nature Switzerland AG 2022
K. Matsui et al. (Eds.): DCAI 2021, LNNS 327, pp. 119–128, 2022.
https://doi.org/10.1007/978-3-030-86261-9_12

endeavor to develop a predictive model capable screening a patient to find out if he/she has any risk of developing Coronary Heart Disease (CHD) in the next 10 years, using Data Mining (DM) and following the CRISP-DM (Cross Industry Standard Process for Data Mining) methodology. DM is a process used to transform raw data into useful knowledge. This transformation focuses on apply specific Machine Learning (ML) in order to extract meaningful patterns from the data [12]. ML is an application of Artificial Intelligence and studying its algorithms, medical systems can be supplied with the capacity to mechanically learn and improve from experience [9]. In order to detail the work carried out in this study, the present paper is structured in four distinct sections. Next, the methodology adopted in this study is described as well as each one of its stages. Afterwards, the obtained results are presented and discussed. Finally, the last sections, contains the main conclusion drawn and some future work is outlined.

2 Methodology

The DM process presented in this study followed the CRISP-DM methodology, as previously mentioned. This methodology provides a framework on how to address DM problems, consisting of six phases: Business Understanding, Data Understanding, Modeling, Evaluation and Deployment [11]. The software used to conduct this study was Rapid Miner. It is a data science software platform that provides an integrated environment for Data Preparation, ML, and Deep Learning capable of functioning as an advanced analytical solution. This software uses template-based frameworks and the minimal need to write code leads to error reducing [13].

2.1 Business Understanding

Heart disease is a major cause of morbidity and mortality globally. However, it is also true that conditions such as heart attacks are highly preventable. Simple lifestyle modifications coupled with early treatment greatly improve its prognosis thus saving lives and allowing a better quality of life. The problem lies in the difficulty in identifying high risk patients due to the high variety of symptoms and behaviors that may cause heart diseases. As such the need for a screening tool capable of identifying patterns and developing clinical awareness regarding heart problems is apparent. Developing an accurate prediction mechanism capable of inferring, based on data provided for each patient, if said patient has a risk of developing any kind of CHD in the next 10 years is the objective of this study.

2.2 Data Understanding

The data used for this project was collected in Framingham, Massachusetts, and is related to the identification of CHD risk factors and the identification of patients at risk of developing CHD within next 10 years [2]. The dataset contains information on 4238 patients and 15 features, each representing a potential risk factor. Among these there are demographic, behavioral, and medical risk factors. These attributes are presented in Table 1, where the attribute in bold represents the target.

Table 1. Attributes of the dataset

Attribute	Description	Type
TenYearCHD	10 year risk of coronary heart disease (CHD)	Discrete
sex	Gender of the patient	Discrete
education	Education of the patient	Discrete
currentSmoker	Whether the patient is a current smoker	Discrete
BPMeds	Whether the patient was on blood pressure medication	Discrete
prevalentStroke	Whether the patient had a stroke previously	Discrete
prevalentHyp	Whether the patient is hypertensive	Discrete
diabetes	Whether the patient has diabetes	Discrete
age	Age of the patient in years	Continuous
cigsPerDay	The number of cigarettes that the patient smokes on average in one day	Continuous
totChol	Total cholesterol level	Continuous
sysBP	Systolic blood pressure	Continuous
diaBP	Diastolic blood pressure	Continuous
BMI	Body Mass Index	Continuous
heaRate	Heart rate of the patient	Continuous
glucose	Whether the patient has diabetes	Continuous

These attributes represent a variety of characteristics of patients. Through the information provided, the predictive model will attempt to infer whether the person will have CHD within the next 10 years. As such, the target is the TenYearCHD attribute, which is "0" if the patient is not at risk of developing CHD in the next 10 years and "1" if the patient is at risk of developing CHD in the next 10 years.

2.3 Data Preparation

The first steps taken were data cleaning and pre-processing. Missing and duplicate variables were removed as these can highly affect the performance of the ML algorithms and the way it is presented in the final dataset may influence the models' training. The work presented strives to assure a simple and easy way to understand the dataset, making sure no highly complicated features or with a low value for the training process are used. Regarding missing values, the glucose feature presented a loss of 9.16%. Other features such as Education and *BPMeds* presented respectively, 2.48% and 1.25% of missing values. In a first approach, to facilitate the data treatment and not erase a large number of instances, the glucose attribute was dropped. Afterwards, to obtain a dataset with more concise information, following the medical guidelines for attributes such as *age*, *total cholesterol*, *BMI*, *blood pressure* and *heart-rate* values, discretization according to the cutoff values was conducted [10]. First, the research on whether age is a risk factor concluded that the risk increases with age group, ranging from low at 20 to 39 years and normal from 40 until 59. However, people with ages above 60 presented an increasingly greater risk of developing CHD [6]. These groups were to discretize the age attribute. To better understand and to take more advantage of the cholesterol-related at-tribute it was transformed into a new one, *Col_strat*. This new attribute divided, according to the cholesterol value and the risk it represented, the instances in three groups: Healthy (where the value was below 200), Borderline high (between 200 and 239), and finally High risk (above 240) [4]. Then, the *BMI* attribute was also transformed according to the metrics used in medical procedures presenting four values: underweight, healthy weight, overweight and lastly obesity when the value was 30 and above, being created the *BMI_strat* attribute [14]. To better deal with the systolic and diastolic blood pressure attributes, as done with the attribute before and following medical guidelines, it was created *SisDia_Strat*. This is a polynomial attribute ranging from 0 to 7 where each value represented a higher risk for heart diseases and general health problems. The heart-rate was addressed in the same way and medical guidelines followed by medicine practitioners were responsible for dictating the thresholds for each range of values [3,16]. The *cigsPerDay* attribute was transformed into *cigsPacksyear* which is a unit that is more widely used in works related to medicine. Following this, the threshold for ranges was also defined transforming it into a polynomial value [5]. With all the attributes defined, the normalization of the values in each was performed. The aim of this normalization was to change the values of numeric columns in the dataset to a common scale, without distorting differences in the ranges of values. Finally, the last step of this phase was the process of oversampling. During the analysis of the dataset, it was found that there was some imbalance when checking the value distribution for the label, *TenYearCHD*. The positive values where very few, 644, while the negative values were 3594. Training a model with an imbalanced data set may result in a biased predictive model towards one class that overshadows the other, thus achieving high accuracy but poor sensitivity or specificity. To address this problem the dataset was synthetically balanced, using the Synthetic Minority Oversampling Technique (SMOTE), which selects a minority class instance randomly and finds its k nearest neighbours and selecting one randomly. The synthetic instances are generated as a combination of the two chosen instances [7]. After using this technique and applying all the other data preparation steps, the resultant dataset presented 3392 negative cases and 3392 positive cases.

2.4 Modeling

In the modeling process, six classification algorithms or Data Mining Techniques (DMT) were used: Decision Tree (DT), Logistic Regression (LR), Random Forest (RF), Naïve Bayes (NB), k-Nearest Neighbors (k-NN), and Support Vector Machine (SVM). Two sampling methods were selected in order to divide the dataset into the training and testing set, namely Split Validation (SV) and Cross-Validation (CV). The SV technique splits the dataset according to the defined percentage values. In this case, it was used 70% for the training set and 30% for the testing set. On the other hand, with CV the dataset is divided into k-folds using each fold as test set. In the end, it is the mean of the error in every fold that is used to calculate the final error value. In this study, it was used $k = 10$. Both sampling methods were used separately, with the additional intention of comparing these methods performance for each. It is expected that Cross-Validation brings forth better performance and more accurate results since it uses all data for training and testing.

Due to the medical context of the problem at hand, a RapidMiner Metacost operator was used to train the model so that false negatives could be further penalized. The reason for this is that when dealing with human lives, missing a diagnosis of a condition may lead to death, as such it is expected to further diminish such occurrences with this operator. The metacost operator makes the model cost-sensitive by using a cost matrix specified and attributing different weights to the solution possibilities.

Several scenarios were created using feature selection methods to study the influence of different attributes to the prediction of the target variable. To select the best features the first approach was to use a correlation matrix to understand the interaction between attributes and between each attribute and the label. As an example, *cigsPerDay* and *currentSmoker* present very high correlation, 0.77, making the presence of both in the dataset unnecessary. However, the correlation between the label and age attribute, or even with the *sysBP* attribute, is high, indicating the high influence these two have on the label and making them good predictors.

Also, after researching medical guidelines it was possible to ascertain to some degree the most influential features that could cause CHD. This way it was also possible to define a few important attributes that should be present in the dataset, even if at first, they did not present high correlation value with the label.

Finally, the Boruta Feature Selection technique was used, which is a feature selection package in R. It is a wrapper method built around the RF classification algorithm. It attempts to acquire all the important features in a dataset concerning the label by training a classifier on the dataset to obtain the weight for each of the features.

Utilizing the different techniques above exposed, a total of five scenarios were developed. All these scenarios used the same data preparation, namely missing values treatment, discretization, normalization, and oversampling, with exception of the first scenario which did not applied the normalization step.

In the first (S1) and the second (S2) scenarios the features were selected with help of the medical guidelines researched. The features selected were *Age_strat*, *BMI_strat*, *cigPacksYear*, *ColStrat*, *HeartRate_strat*, *Sis_Dia_strat*, *BPMeds*, *Diabetes*, *Male*, *Education*, *PrevHyp* and *prevStroke*.

As mentioned before, the difference between these two scenarios relies on the application of normalization, which has the goal of comparing the influence of this step in the performance of the final model. The third scenario (S3) the feature selection method applied also implemented the knowledge acquired from the correlation matrix. The features selected for the data set were *Age_strat*, *BMI_strat*, *cigPacksYear*, *ColStrat*, *HeartRate_strat*, *Sis_Dia_strat*, *Diabetes*, and *Male*. The fourth scenario (S4) used the Boruta selection method to select the most important data attributes. The ones that were part of the final selection were *Age*, *total cholesterol*, *systolic blood pressure*, *diastolic blood pressure*, *BMI*, *heart rate*, and *blood glucose*. Finally, the last scenario (S5), was created using a combination of all the feature selection methods mentioned before an it was composed by *Age*, *BMI*, *cigPacksYear*, *total cholesterol*, *HeartRate*, *Sis_Dia_strat*, *Diabetes*, and *Male*.

2.5 Evaluation

To evaluate the performance of each Data Mining Model (DMM), different metrics were considered, namely accuracy, Area under Curve (AUC), and F1-Score. The reason for choosing these metrics focuses on an all-round performance capture regarding True Positives (TP), True Negatives (TN), False Positives (FP) and False Negatives (FN) retrieved from the resulting confusion matrix [8].

Accuracy is the ratio between correctly predicted observations and the total value of observations, being calculated through Eq. 1.

The AUC provides an aggregate measure of performance across all possible classification thresholds. It gives the probability that the model ranks a random positive example more highly than a random negative example, representing the degree or measure of separability. It tells how much the model is capable of distinguishing between classes [16].

F1-Score, calculated by Eq. 2, uses both precision and recall and achieves a more realistic measure of a test's performance taking both FP and FN into account.

Precision is the ratio of correctly predicted positive observations to the total predicted positive observations and is calculated by Eq. 3.

Recall is the ability of a test to correctly identify positive results to get the true positive rate it is calculated by Eq. 4.

$$Accuracy = \frac{TP + TN}{TP + FP + FN + TN} \tag{1}$$

$$F1Score = \frac{2 \times Precison \times Recall}{Precision + Recall} \tag{2}$$

$$Precision = \frac{TP}{TP + FP} \tag{3}$$

$$Recall = \frac{TP}{TP + FN} \tag{4}$$

3 Results and Discussion

After the implementation of the methods developed, the information about the performance values for each classifier is presented in Tables 2, 3, 4, 5 and 6 for S1, S2, S3, S4 and S5 respectively, using both CV and SV.

By analyzing these tables from S1 through S5, it is possible to conclude that the CV sampling method presented better values across most scenarios and evaluation metrics. Overall, the DT algorithm accomplished the best performance across all evaluation metrics with S5 and using the CV sampling method with an accuracy of 0.884, an AUC of 0.942 and a F1-Score of 0.881. This high accuracy means that the model is capable of predicting the occurrence of CHD within 10 years efficiently. The high AUC and F1-Score also show that the model has a high true positive rate and is thus sensitive to predict if a patient has a high risk of developing CHD. Overall, the second-best performing algorithm was k-NN, except for S4 in which RF was the second-best classifier. It is also interesting to note that S1 obtained worst results in almost every metrics than S2, being the only difference between them the use of normalization by S2. In this sense, it is proven that the normalization of the data is a valuable step in order to optimize the learning process of the models. Regarding S3 and S4, there were no substantial differences in the performance. These scenarios did not obtain the best neither the worst results, but between them, S3 obtained better results with all algorithms (despite different sampling methods), except for the k-NN algorithm. The worst performance was obtained by NB algorithm in S1 when using CV. In addition, when analyzing the results for all the scenarios, the NB algorithm obtained the worst values in each one. Rating the scenarios from worst to best, it is possible to conclude that S1 obtained the worst results, followed by S4, S3, S2 and S5, which contained the best result.

Table 2. Results obtain for S1 for each DMT, sampling method and evaluation metric

DMT	Accuracy		AUC		F1-Score	
	CV	SV	CV	SV	CV	SV
DT	0.759	0.721	0.8333	0.794	0.773	0.728
LR	0.642	0.669	0.701	0.731	0.699	0.686
RF	0.660	0.707	0.743	0.799	0.692	0.722
NB	0.597	0.654	0.687	0.706	0.450	0.651
KNN	0.733	0.716	0.818	0.790	0.718	0.694
SVM	0.632	0.665	0.683	0.703	0.634	0.597

Table 3. Results obtain for S2 for each DMT, sampling method and evaluation metric

DMT	Accuracy		AUC		F1-Score	
	CV	SV	CV	SV	CV	SV
DT	0.871	0.854	0.932	0.913	0.866	0.844
LR	0.663	0.678	0.715	0.743	0.677	0.691
RF	0.833	0.830	0.916	0.915	0.834	0.831
NB	0.653	0.663	0.707	0.723	0.653	0.658
KNN	0.798	0.796	0.832	0.826	0.801	0.794
SVM	0.632	0.649	0.687	0.701	0.646	0.635

Table 4. Results obtain for S3 for each DMT, sampling method and evaluation metric

DMT	Accuracy		AUC		F1-Score	
	CV	SV	CV	SV	CV	SV
DT	0.877	0.824	0.927	0.903	0.869	0.825
LR	0.652	0.677	0.701	0.723	0.667	0.693
RF	0.769	0.806	0.889	0.898	0.791	0.816
NB	0.643	0.662	0.696	0.716	0.660	0.682
KNN	0.782	0.791	0.853	0.807	0.772	0.811
SVM	0.659	0.731	0.704	0.813	0.679	0.765

Table 5. Results obtain for S4 for each DMT, sampling method and evaluation metric

DMT	Accuracy		AUC		F1-Score	
	CV	SV	CV	SV	CV	SV
DT	0.789	0.806	0.868	0.871	0.798	0.811
LR	0.665	0.668	0.716	0.706	0.679	0.685
RF	0.779	0.768	0.871	0.845	0.789	0.788
NB	0.630	0.653	0.704	0.723	0.566	0.598
KNN	0.705	0.815	0.812	0.850	0.762	0.837
SVM	0.666	0.684	0.713	0.730	0.681	0.706

Table 6. Results obtain for S5 for each DMT, sampling method and evaluation metric

DMT	Accuracy		AUC		F1-Score	
	CV	SV	CV	SV	CV	SV
DT	0.884	0.828	0.942	0.905	0.881	0.829
LR	0.671	0.677	0.723	0.723	0.689	0.693
RF	0.669	0.806	0.819	0.898	0.735	0.816
NB	0.668	0.662	0.715	0.716	0.686	0.682
KNN	0.796	0.791	0.876	0.807	0.820	0.811
SVM	0.681	0.668	0.736	0.720	0.701	0.680

4 Conclusion

The DM process plays an important role in the health sector, as it improves the quality of life of patients and can also save lives. This paper shows the role of DM in predicting CHD through the analysis of different risk factors. This study primarily consisted of the implementation of DM techniques to predict the possibility of a patient developing CHD within the next 10 years. The best learning model was obtained by the DT algorithm, with the application of the CV technique, using only the *Age*, *BMI*, *cigPacksYear*, *total cholesterol*, *HeartRate*, *Sis_Dia_strat*, *Diabetes*, and *Male* features (S5), obtaining an accuracy of 0.884, an AUC of 0.942 and a F1-Score of 0.881. Since the goal of this predictive model was directly related to the patient's health, FN present a serious problem as they may cause medical misguidance when dealing with a patient. After refining, enhancing and validating this work, the developed models have the potential to be integrated in decision making systems used by medical practitioners to ascertain the risk of their patients developing CHD in the coming years. As future work it would be important to collect more cases of the positive class, as most of these values in the used dataset were artificially synthesized using SMOTE and, consequently, they may not be a true representation of the actual population. This data collection is needed to endeavor in developing and assuring a more reliable predictive model and much more accurate screening tool.

Acknowledgements. This work has been supported FCT—Fundação para a Ciência e Tecnologia (Portugal) within the Project Scope: UIDB /00319/2020.

References

1. Heart disease facts (2020). https://www.cdc.gov/heartdisease/facts.html
2. Ajmera, A.: Framingham heart study dataset (2021). https://www.kaggle.com/amanajmera1/framingham-heart-study-dataset
3. Ettehad, D., et al.: Blood pressure lowering for prevention of cardiovascular disease and death: a systematic review and meta-analysis. The Lancet **387**(10022), 957–967 (2016)
4. Grundy, S.M., et al.: 2018 aha/acc/aacvpr/aapa/abc/acpm/ada/ags/apha/aspc/nla/pcna guideline on the management of blood cholesterol: executive summary: a report of the American college of cardiology/American heart association task force on clinical practice guidelines. J. Am. Coll. Cardiol. **73**(24), 3168–3209 (2019)
5. Guerrero, J., Segarra, M., Chorro, J., Bataller, M., Rosado, A., Espi, J.: Early effect of the suppression of the smoking habit on the heart rate variability. In: Computers in Cardiology, pp. 425–428. IEEE (2002)
6. Mozaffarian, D., et al.: Heart disease and stroke statistics-2015 update: a report from the American heart association. Circulation **131**(4), e29–e322 (2015)
7. Neto, C., Brito, M., Lopes, V., Peixoto, H., Abelha, A., Machado, J.: Application of data mining for the prediction of mortality and occurrence of complications for gastric cancer patients. Entropy **21**(12), 1163 (2019)
8. Neto, C., Peixoto, H., Abelha, V., Abelha, A., Machado, J.: Knowledge discovery from surgical waiting lists. Procedia Comput. Sci. **121**, 1104–1111 (2017)
9. Nithya, B., Ilango, V.: Predictive analytics in health care using machine learning tools and techniques. In: 2017 International Conference on Intelligent Computing and Control Systems (ICICCS), pp. 492–499. IEEE (2017)
10. Perk, J., et al.: European guidelines on cardiovascular disease prevention in clinical practice (version 2012). The fifth joint task force of the European society of cardiology and other societies on cardiovascular disease prevention in clinical practice (constituted by representatives of nine societies and by invited experts). Giornale italiano di cardiologia (2006) **14**(5), 328–392 (2013)
11. Schäfer, F., Zeiselmair, C., Becker, J., Otten, H.: Synthesizing crisp-DM and quality management: a data mining approach for production processes. In: 2018 IEEE International Conference on Technology Management, Operations and Decisions (ICTMOD), pp. 190–195. IEEE (2018)

12. Shadabi, F., Sharma, D.: Artificial intelligence and data mining techniques in medicine-success stories. In: 2008 International Conference on BioMedical Engineering and Informatics, vol. 1, pp. 235–239. IEEE (2008)
13. Sharma, P., Singh, D., Singh, A.: Classification algorithms on a large continuous random dataset using rapid miner tool. In: 2015 2nd International Conference on Electronics and Communication Systems (ICECS), pp. 704–709. IEEE (2015)
14. Shashikant, R., Chetankumar, P.: Effect of obesity on heart rate variability among obese middle-aged individuals. In: 2019 International Conference on Advances in Computing, Communication and Control (ICAC3), pp. 1–5. IEEE (2019)
15. Wiles, R., Kinmonth, A.L.: Patients' understandings of heart attack: implications for prevention of recurrence. Patient Educ. Counseling **44**(2), 161–169 (2001)
16. Williams, B., et al.: 2018 esc/esh guidelines for the management of arterial hypertension: the task force for the management of arterial hypertension of the European society of cardiology (esc) and the European society of hypertension (esh). Eur. Heart J. **39**(33), 3021–3104 (2018)

Soft-Sensors for Monitoring *B. Thuringiensis* Bioproduction

C. E. Robles Rodriguez[1], J. Abboud[2], N. Abdelmalek[3], S. Rouis[4], N. Bensaid[3], M. Kallassy[5], J. Cescut[2], L. Fillaudeau[1], and C. A. Aceves Lara[1](✉)

[1] Toulouse Biotechnology Institute, Bio & Chemical Engineering, and the CNRS, UMR5504, Université de Toulouse, UPS, INSA, INP, TBI F31077, the INRA, UMR792, 135 avenue de Rangueil, 31077 Toulouse, France
{roblesro,luc.fillaudeau,aceves}@insa-toulouse.fr

[2] Toulouse White Biotechnology (UMS INRAE/INSA/CNRS), 135 Avenue de Rangueil, 31077 Toulouse CEDEX 04, France
{joanna.abboud-1,Julien.Cescut}@inrae.fr

[3] Laboratoires Pharmaceutiques MédiS, Nabeul, Tunisia
nadia.bensaid@labomedis.com

[4] Centre of Biotechnology of Sfax, B.P 1177, 3018 Sfax, Tunisia
souad.rouis@cbs.rnrt.tn

[5] Faculty of Sciences Saint, Joseph University, Riad El Solh, Beirut, Lebanon
mireille.kallassy@usj.edu.lb

Abstract. One of the main challenges in ferentation cultures is the monitoring of key variables that can indicate the performance and help in the optimization of the bioprocess. On-line estimation could be a challenging task when an accurate process model is not available. Support Vector Machine (SVM) is an attractive and relatively simple method that can be used as an alternative to predict key variables by using several physical parameters measured online. In this paper, we show the application of SVM to the production of protein by *B. thuringiensis*. The soft-sensor was trained and validated with independent data sets of batch fermentations evaluating the impact of different predictor variables and kernels. Results show that protein production can be predicted only using online measurements.

Keywords: Soft-sensors · Monitoring · Support vector machine · Proteins and Spores optimization · B. thuringiensis bioproduction

1 Introduction

B. thuringiensis is a facultative anaerobic gram-positive sporulating bacterium, frequently used in the production of some biopesticides and as a source of genes for transgenic expression in plants [1]. It is usually found in different environments, among which soil, settled dust, insects, water, and others have been identified [2]. *B. thuringiensis* has been shown to be toxic to various organisms such as lepidopterans, coleopterans, dipterans, or nematodes, but is considered safe for mammals. Thus, the products based on *B. thuringiensis* (Bt) provide effective and environmentally benign control of

© The Author(s), under exclusive license to Springer Nature Switzerland AG 2022
K. Matsui et al. (Eds.): DCAI 2021, LNNS 327, pp. 129–136, 2022.
https://doi.org/10.1007/978-3-030-86261-9_13

several insects in agricultural, forestry and disease-vector applications [3]. This insecticidal activity is mainly due to the production of some intracellular inclusions (called δ-endotoxins) during the post-exponential phase of *B. thuringiensis* cells. Most of the biopesticides distributed in the world are principally based on *B. kurstaki* HD1 strain. However, a recent strain, identified as *B. kurstaki* Lip, has been isolated and described to be more efficient than HD1 [4]. Therefore, this last strain will be studied in this work.

The monitoring of certain variables of *B. thuringiensis* culture is complicated due to several changes of cell physiology during growth, which will hamper the optimization of the fermentation. Usually, it is difficult to measure online substrate, biomass, and product concentrations in the bio-process. The so-called "soft-sensors" are an alternative for online estimation. Soft sensors are software based sophisticated monitoring systems, which can relate the infrequently measured process variables with the easily measured [5]. In this way, these soft-sensors assist in obtaining a real-time prediction of the unmeasured variables [6].

Several software sensors have been proposed for fermentation such as Support Vector Machine (SVM) [7] and Decision Trees DT [8]. A recent review paper [9] has shown the use of methods like neural networks, fuzzy logic, SVM, genetic algorithms and probabilistic latent variable models in fermentation, where the authors highlighted that SVM has become an indispensable method to measure internal variables, especially when small amount of data exists [10]. Furthermore, SVM shares many of its features with the artificial neural networks, but it proposes some additional characteristics [5]. It has good generalization ability of the regression function, robustness of the solution, and sparseness of the regression [6].

In this context, this work proposes some SVM soft-sensors to monitoring the production of a protein by *B. thuringiensis* with the purpose of showing the application of SVM for microorganisms with physiology changes during fermentation. The remainder of the paper is as follows: the experiments are presented in Sect. 2, the methodology of soft-sensors is presented in Sect. 3. Section 4 holds the results of the soft-sensor with the training and validation data. Finally, Sect. 5 reports the conclusions and some perspectives of this work.

2 Material and Methods

2.1 Organism and Culture Media

B. thuringiensis Lip is a Lebanese strain [4]. Luria broth (LB) was used for inoculum production, whereas, a semi-synthetic medium (SSM). For the SSM, concentrated glucose (Sol 2) and all salts solutions (Sol 3, 4, 5) were prepared and sterilized separately and added before inoculation to the rest of medium (Sol 1) previously sterilized.

2.2 Fermentation Conditions

Several fermentations were conducted in batch mode at 30 °C in a 3 L Biostat B plus fermenter (Sartorius; Germany) containing 1.8 L of the SSM medium and with continuous regulation of pH at 6.8 using 1 M H3SO4 and 3 M NaOH. Dissolved oxygen was continuously monitored by an optical oxygen sensor and maintained at 25% pO2sat with a constant aeration rate (0.18min/L) and variable stirring. Foaming was controlled by the use of an antifoam (Emultrol DFM DV-14 FG), through the fermentation process.

2.3 Total Cell and Spores Count

The follow up of the Bt culture was performed by estimating total cell counts (TC) and spore counts (SC) by plate counts. To determine TC, the withdrawn samples were serially diluted, spread on LB plates and incubated at 30 °C for 16–18 h. As for SC, the appropriately diluted samples were heated at 85 °C for 15 min and cooled for 5 min before spreading onto LB plates and then incubated at 30 °C.

2.4 Dry Matter

Biomass dry weight (DW) concentration was determined by filtering a known amount of sample and differential weighing of the filter before and after oven drying at 70 °C.

2.5 Quantification of Delta Endotoxins Production

The concentration of the delta-endotoxins was determined by Bradford assay using bovine serum albumin (BSA) as a protein standard. Samples were measured after 10 min at 595 nm. The obtained value was the average of three measures of the same sample.

2.6 Sugar Analysis

The sugar concentrations were determined using HPLC-UV with a (Ultimate 3000 RSLC/MWD/RI/CAD). A mobile phase of 5 mM H2SO4 with a flow rate of 0.6 mL/min was used. The mobile phase was filtered and degassed through a 0.2 μm cellulose nitrate membrane. The samples and standards were also filtered before injection into the HPLC.

3 Support Vector Machine

Support vector machine (SVM) is a kernel-based tool for solving pattern recognition and regression problems. The idea of SVM relies on mapping the input data or features $x \in R^{M \times N}$ into a nonlinear space in order to predict a desired vector of outputs $y \in R^M$ [9]. The objective of the regression analysis is to determine a function that predicts accurately the desired outputs y in the form

$$y = w^T \varphi(x) + b \tag{1}$$

where $\varphi(x)$ is the nonlinear mapping of the inputs x into a high dimensional feature space. The vector w represents the support vectors and b is the bias term. The determination of the support vectors is performed solving the following optimization problem:

$$\min_{w,\, \xi,\, \xi^*} J = \frac{1}{2} w^2 + C \sum\nolimits_{i=i}^{N} (\xi_\iota + \xi_i^*) \tag{2}$$

$$\text{Subject to} \begin{cases} d_i \leq \varepsilon + \xi_i \\ -d_i \leq \varepsilon + \xi_i^* \\ \xi_i, \ \xi_i^* \geq 0 \end{cases} \quad (3)$$

with $d_i = y_i - w^T \varphi(x_i) - b$.

In Eq. (2) the first term is the regularized term, and the second term is the empirical error (risk) measured by the insensitive ε-loss function enabling to use less data points to represent the decision function given by Eq. (1). The variables ξ_i and $\xi_i{}^*$ are the slack variables that measure the deviation of the support vectors from the boundaries of the ε -zone and determine how strictly the model fits the data. The constant C is the regularization constant. It determines the trade-off between the empirical risk and the regularized term. The term ε is called the tube size and it is equivalent to the approximation accuracy placed on the training data points. Both parameters will determine the efficiency of the estimation [5].

In order to simplify the dual minimization problem, Lagrange multipliers are introduced as follows:

$$L = J - \sum\nolimits_{i=1}^{N} \alpha_i\{\xi_i + \varepsilon - d_i\} - \sum\nolimits_{i=1}^{N} \alpha_i^*\{\xi_i^* + \varepsilon + d_i\} - \sum\nolimits_{i=1}^{N} (\eta_i\xi_i - \eta_i^*\xi_i^*) \quad (4)$$

where the parameters α_i, α_i^*, η_i, and η_i^* are the Lagrange multipliers. According to the Karush-Kuhn-Tucker (KKT) of quadratic programming, the dual equation that can be obtained [6] is:

$$min_{\alpha,\,\alpha^*} W = \frac{1}{2}\sum\nolimits_{i,j=i}^{N} (\alpha_\iota - \alpha_i^*).\left(\alpha_j - \alpha_j^*\right)K\left(x_i,\ x_j\right) + \varepsilon \sum\nolimits_{i=i}^{N} (\alpha_\iota + \alpha_i^*) - \sum\nolimits_{i=i}^{N} (\alpha_\iota - \alpha_i^*)y_i \quad (5)$$

$$\text{Subject to} \begin{cases} \sum_{i=i}^{N} (\alpha_\iota - \alpha_i^*) \\ 0 \leq \alpha_\iota,\ \alpha_i^* \leq C; \quad i = 1, 2, \ldots, N \end{cases} \quad (6)$$

Therefore, the final regression function given in Eq. (1) is rewritten as

$$y = \sum\nolimits_{i=i}^{N} (\alpha_\iota - \alpha_i^*)K\left(x_i,\ x_j\right) + b \quad (7)$$

where $K\left(x_i,\ x_j\right) = \varphi(x_i)\varphi\left(x_j\right)$ is the kernel function that corresponds to any symmetric function satisfying the Mercer's condition. The most typical examples of the kernel functions are the polynomial kernels and the Radial Basis Function (Gaussian) kernels whose mathematical representation is

$$K\left(x_i,\ x_j\right) = [(x.x_i) + 1]^d \quad (8)$$

$$K\left(x_i,\ x_j\right) = \exp\left(-\left(x - x_i^2/2\sigma^2\right)\right) \quad (9)$$

where d is the order of the polynomial and σ represents the width of the RBF. The most used Kernel function is the Radial Basis Function (RBF) because it can classify multi-dimensional data. As the prediction depends on the type of kernel and their involved parameter, in this work we will compare three types of kernels and assess their prediction capability for protein production.

4 Results

The support vector machine algorithm has been applied to predict Protein production by *B. thuringiensis*. The available data correspond to ten batch experiments performed with three different strains. The available online measurements of the experiments were partial oxygen pressure (pO2), pH, agitation (Agit) and aeration (Aer); whereas the off-line measurements regarded optical density (OD), Biomass (Bio), Glucose (Glc), flora (Fl), and spores (Sp). The nine variables have been used as inputs to SVM. Additionally, the logarithms of Flora and Spores have been added as inputs as well as the number of strains. Before performing SVM, the inputs were ranked according to the Neighborhood Component Analysis in order to guide feature selection. The most important variables corresponded to pO2, strain number, Biomass and Spores. The least relevant inputs for protein production were the Flora and the logarithm of the flora. As the kernel type is a limiting factor in the performance of SVM, we have explored three kernels to assess their application.

Table 1. Combinations to determine the best model for Protein prediction.

	Model							
	1	2	3	4	5	6	7	8
	Gaussian Kernel							
RMSEt	189.35	64.96	66.18	57.41	65.18	66.22	61.26	66.42
RMSEv	207.50	174.57	133.62	152.31	129.26	126.65	165.01	156.05
	Quadratic Kernel							
RMSEt	200.74	128.01	103.62	94.75	50.45	53.40	191.29	185.30
RMSEv	194.30	316.80	166.32	175.14	138.05	151.54	228.61	235.21
	Linear Kernel							
RMSEt	218.98	208.41	186.31	175.43	122.89	134.42	145.46	148.65
RMSEv	199.95	184.66	174.75	189.71	218.33	162.90	168.17	172.29
Predictors								
OD	0	0	0	1	1	1	1	0
Bio	0	1	1	1	1	1	0	0
Glc	0	0	1	1	1	1	0	0
pO2	1	1	1	1	1	1	1	1
Agit	0	0	0	0	1	1	1	1
pH	0	0	0	1	1	1	1	1
Aer	0	0	0	0	1	1	1	1
Fl	0	0	0	0	0	1	0	0
Sp	0	0	1	1	1	1	0	0
Log(Fl)	0	0	0	0	0	1	0	0
Log(Sp)	0	1	1	1	1	1	0	0
Strain	1	1	1	1	1	1	1	1

The training of the SVM has been performed in MATLAB using the fitrsvm command from the regression learner application. Seven data sets were used for training with an 8 k-fold cross validation. The training sets comprised batch experiments: 2, 3, 4, 5, 7, 8, 10. The remaining three data sets consisting on batches 1, 6 and 9 one per strain number) were considered for the validation of the soft-sensor.

Several combinations of input variables have been explored with three different kernels: linear, quadratic and gaussian. The eight combinations have been assessed via the RMSE. The performance analysis of the SVM models are reported in Table 1 where RMSEt and RMSEv correspond to the errors for the training and validation, respectively. Table 1 indicates that the best prediction for the training data is achieved by Model 5 with a quadratic kernel. Furthermore, this model provides a good between the training and validation sets. However, it needs 10 predictor variables. The prediction with Model 5 is displayed in Fig. 1 where it can be appreciated that most of the data points are well predicted and some points could be removed to improve the performance since they seem to be outliers in the prediction. For instance, see the two last points of Batch 10, which are at a very low value. Furthermore, it is worth noting that the predictions are well adapted for the different types of strains producing lower titers of protein (*i.e.* batch 8, 9,10).

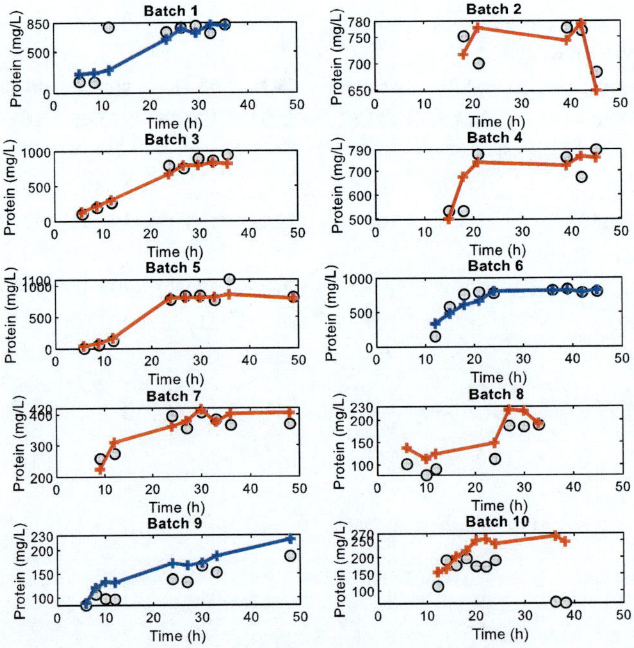

Fig. 1. Results of the SVM model 5 with the training sets (red lines) and the validation sets (blue lines). The dots represent the experimental data points.

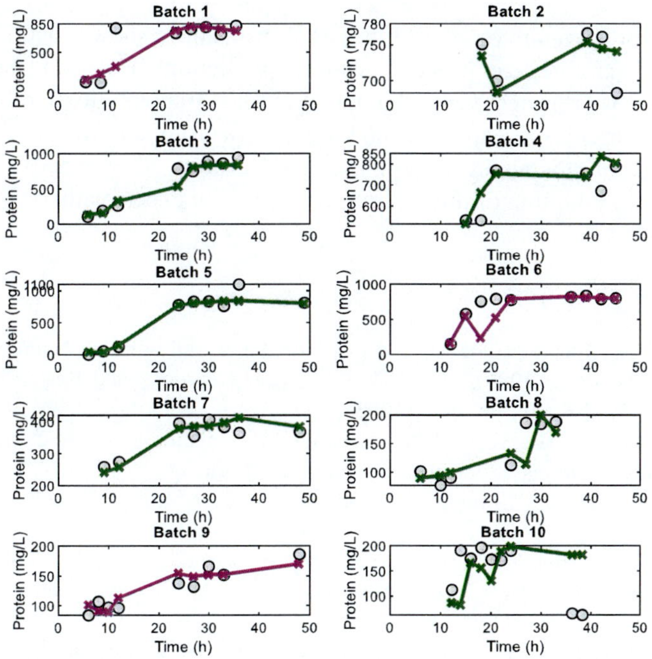

Fig. 2. Prediction of protein production with the SVM model 8 and gaussian kernel against the training sets (green lines) and the validation sets (purple lines). The dots represent the experimental data points.

Although model 5 provides good prediction, one of the goals of the development of soft-sensors for fermentation is their ability to be used at different conditions and scales. In this context, we have explored models only using online measurements. This is the case of model 8 and model 7 assuming that OD can sometimes be measured online. It is worth noting that model 8 which uses a gaussian kernel also provides a good compromise between the training and validation sets. Additionally, the RMSE values are similar to the results obtained by model 5. The prediction of model 8 is shown in Fig. 2. It is important to note that the one if the most important input is the strain number which allows to predict correctly among the different cultures. Model 8 is chosen as the best one because it presents higher online applicability, which is one of the objectives of the proposition of an online software sensor. These results highlight the fact that SVM can represent accurately the nonlinearities of the process.

5 Conclusions

This work introduced a soft-sensor based on SVM for the prediction of protein by different strains of *B. thuringiensis*. The SVM algorithm was successfully implemented to generate an online monitoring of the concentration of protein production, which is normally measured off-line. The results proved that SVM is an attractive for monitoring providing a good tradeoff between the quality of the approximation of the given data and the complexity of the approximating function. Results have shown that diverse combinations of input variables can produce accurate predictions. However, the model only using online data is preferred due to the potential for extrapolability to other conditions and, especially, for industrial application. Future work will focus on increasing the quality of the prediction, the application of the soft-sensor in other experimental working conditions and the coupling of the soft-sensor for in-situ experiments.

References

1. Schnepf, E.: Bacillus thuringiensis and its pesticidal crystal proteins. Microbiol. Mol. Biol. Rev. **62**(3), 775–806 (1998). Chen, W.-K.: Linear Networks and Systems (Book style), pp. 123–135. Wadsworth, Belmont (1993)
2. Iriarte, J., Porcar, M., Lecadet, M.-M., Caballero, P.: Isolation and characterization of bacillus thuringiensis strains from aquatic environments in Spain. Curr. Microbiol. 40, 402–408 (2000). Smith, B.: An approach to graphs of linear forms (Unpublished work style). unpublished
3. Rowe, G.E., Margaritis, A.: Bioprocess design and economic analysis for the commercial production of environmentally friendly bioinsecticides from bacillus thuringiensis HD-1 kurstaki. Biotechnol. Bioeng. **86**(4) (2004)
4. El Khoury, M., Azzouz, H., Chavanieu, A., Abdelmalak, N., Chopineau, J., Awad, M.K.: Isolation and characterization of a new Bacillus thuringiensis strain Lip harboring a new cry1Aa gene highly toxic to Ephestia kuehniella (Lepidoptera: Pyralidae) larvae. Arch. Microbiol. **196**(6), 435–444 (2014)
5. Vapnik, V., Golowich, S.E., Smola, A.: Support vector method for function approximation, regression estimation, and signal processing. Annu. Conf. Neural Inf. Process. Syst. 281–287 (1996)
6. Liu, G., Zhou, D., Xu, H., Mei, C.: Model optimization of SVM for a fermentation soft sensor. Expert Syst. Appl. **37**, 2708–2713 (2010)
7. Ou Yang, H.-B., Li, S., Zhang, P., Kong, X.: Model penicillin fermentation by least squares support vector machine with tuning based on amended harmony search. Int. J. Biomath. **08**, 1550037 (2015)
8. Ahmad, M.W., Reynolds, J., Rezgui, Y.: Predictive modelling for solar thermal energy systems: a comparison of support vector regression, random forest, extra trees and regression trees. J. Clean. Prod. **203**, 810–821 (2018)
9. Zhu, X., Rehman, K.U., Wang, B., Shahzad, M.: Modern soft-sensing modeling methods for fermentation processes. Sensors(Switzerland), **20**(6), 1771 (2020).
10. Jianlin, W., Tao, Y.U., Cuiyun, J.I.N.: On-line estimation of biomass in fermentation process using support vector machine. Chinese J. Chem. Eng. **14**, 383–388 (2006)

A Tree-Based Approach to Forecast the Total Nitrogen in Wastewater Treatment Plants

Carlos Faria[1(✉)], Pedro Oliveira[1], Bruno Fernandes[1], Francisco Aguiar[3], Maria Alcina Pereira[2], and Paulo Novais[1]

[1] ALGORITMI Centre, University of Minho, 4710-057 Braga, Portugal
{pedro.jose.oliveira,bruno.fernandes}@algoritmi.uminho.pt, pjon@di.uminho.pt

[2] Centre of Biological Engineering, University of Minho, 4710-057 Braga, Portugal
alcina@deb.uminho.pt

[3] Águas do Norte, Rua Dr. Roberto de Carvalho, n.o 78-90, 4810-284 Guimarães, Portugal
f.aguiar@adp.pt

Abstract. With the increase in the world population, there has been an increase in environmental problems worldwide. One of these problems is the quality of the water, which can cause problems to society's well-being and the environments surrounding it. Wastewater Treatment Plants (WWTPs) emerged to address this problem. It is necessary to pay attention to the different substances present in the waterwaters treated in the WWTPs, as is the case of total nitrogen, which can cause severe damage to the environment. Therefore, this work aims to forecast the total nitrogen in a WWTP by conceiving, tuning and evaluating several Machine Learning (ML) models, particularly the Decision Trees (DTs) and the Random Forest (RF). The best candidate model was a DT-based with an approximate error of 1.6 mg/L. Considering the best candidate model identified, our objective was to extract the rules generated by the model to understand the factors that lead to high values at the level of total nitrogen.

1 Introduction

With the exponential growth of the human population and the need to satisfy essential goods, there has been an increase in multiple problems, such as safety [1] and environmental ones [2]. In fact, concerns about the quality and quantity of freshwater have been substantially increasing, becoming one of the main research topics [3]. Thus, it has become imperative to treat wastewater as it poses a risk to public health, as the appearance of various diseases [4]. To address this need, Wastewater Treatment Plants (WWTPs) have emerged, which perform the clean-up of numerous water streams, where large polluting loads are daily channelled so that the water returns to its natural habitat under normal and environmentally safe conditions [6]. To achieve this purpose, it is necessary to

© The Author(s), under exclusive license to Springer Nature Switzerland AG 2022
K. Matsui et al. (Eds.): DCAI 2021, LNNS 327, pp. 137–147, 2022.
https://doi.org/10.1007/978-3-030-86261-9_14

monitor several effluent parameters to guarantee a high level of water quality. For example, excessive amounts of nitrogen in treated wastewater increase the eutrophication process, which causes a marked degradation of water quality in aquatic environments [15]. Total nitrogen corresponds to the sum of the total nitrogen content determined by the Kjeldahl method (organic and ammoniacal nitrogen), with the nitrogen contained nitrates and nitrites [7].

With this in mind, the goal of this work is to forecast the total nitrogen at the effluent in a WWTP, using Machine Learning (ML) models, namely Decision Trees (DTs) Regression and Random Forest (RF) Regression, which uses the CART algorithm in its construction. The use of tree-based models instead of deep learning took into account the low number of observations in the dataset [8]. In addition, the rules generated in the best candidate model is verified, especially in cases where the total nitrogen value is above the limit defined by Portuguese legislation [9].

This manuscript is structured as follows: the next section summarizes state of the art considering the forecast of total nitrogen in WWTP. The third section presents the materials and methods used, where the datasets are explored and pre-processed. The fourth and fifth sections aim to describe the experiments carried out and discuss the obtained results. The sixth and final section exposes the main conclusions obtained from this study and presents future work.

2 State of the Art

There are already several studies where the prediction of several substances present in treated wastewater is made, among them the total nitrogen [10]–[12]. Bagherzadeh et al. [10] conducted a study where the objective was to predict the total nitrogen in a WWTP effluent. The authors used different feature selection methods to identify the most relevant features with a target feature. The initial dataset contained 800 records between 2015 and 2017. The authors used Artificial Neural Networks (ANNs), RF and Gradient Boosting Machine (GBM) to compare and define four scenarios based on feature selection suggestions. As evaluation metrics, the authors used Root Mean Squared Error (RMSE) and Mean Absolute Error (MAE), and coefficient of determination (R^2). The data were divided into training (96,25%) and test (3.75%) in all models, without cross-validation in the training process. As a result, the tree-based models obtained the best results (RF and GBM) compared to ANN. However, GBM managed to generalize the dataset patterns better and obtained better results than RF.

Another study carried out by Guo et al. [11] used two ML models, namely ANNs and Support Vector Machines (SVMs), to predict the daily concentration of total nitrogen. The data were collected from a WWTP in Ulsan, South Korea, where data on daily water quality and meteorology were used, dividing into training and validation sets. The training data corresponds to a period between January and August and the validation data between September and October. The data were normalized between –1 and 1. Additionally, a Latin-Hypercube One-factor-At-a-Time (LH-OAT) and a Pattern Search algorithm were applied to the sensitivity analysis and optimization of model parameters. Also, the authors used a cross-validation process. The model's performance evaluation methods used by the authors were R^2, Nash-Sutcliffe Efficiency (NSE) and Relative Efficiency Criteria (d_{rel}). The results showed that both models could effectively forecast the daily concentration of total nitrogen in the treated effluent. However, the SVM model showed better performance, with R^2 of 0.46 in the validation phase.

Nourani et al. [12] developed a study using several ML models, such as Adaptative Neuro-Fuzzy Inference System (ANFIS) and SVM, to forecast several substances, including total nitrogen, in a WWTP in Nicosia, Cyprus. The dataset corresponds to the daily measurements, with a period between 2015 and 2016. The parameters selection considers seasonal variations and consists of several parameters observed in the entry and exit of wastewater. For each substance, the authors normalize the data between 0 and 1 and also used cross-validation on the training process. The metrics used to evaluate the models were RMSE and R^2. As a result, the ANFIS model obtained the best results with an R^2 of 0.9571 for total nitrogen. In addition, the authors used different ensemble techniques, such as Neural network ensemble (NNE), which demonstrated better results when compared to a single model, with an R^2 of 0.979.

In general, the studies that used tree-based models do not analyze the rules that can be extracted from the generated tree, and not all models pay attention to the search of the best hyperparameters and the use of cross-validation.

3 Materials and Methods

The materials and methods used in this study are described and detailed in the following lines as well as the metrics to perform ML models. Finally, the ML models used are presented.

3.1 Data Collection

In this study, we use two distinct datasets, one referring to the several indicators present in the affluent and effluent of the WWTP and the second one to the climatological data. Concerning the first dataset, this was provided by a portuguese multi-municipal wastewater company. For the second dataset, we use the Open Weather Map API to collect the WWTP climatological data for the same city. Both datasets correspond to the period between January 2016 and May 2020.

3.2 Data Exploration

The first dataset corresponds to historical data of the substances present int the WWTP affluent and effluent. This dataset consists of 3 features, as described in Table 1, where it contains a total of 1901 records. In the same table, we present

Table 1. Available features in the datasets.

#	Features	Description	Units
		Substances dataset	
1.	Date	Timestamp	Date and time
2.	Indicator_name	Name of the indicator	Text
3.	Value	Value of indicator	Different units
		Climatological dataset	
1.	dt.iso	Timestamp	Date and time
2.	Temp	Temperature	°C
3.	Pressure	Atmospheric pressure	°C
4.	Humidity	Humidity percentage	%
5.	Wind_speed	Wind speed value	m/s
6.	Wind_deg	Wind direction	Degrees
7.	Rain	Rain volume	mm
8.	Cloudiness	Cloudiness percentage	%

some of the characteristics of climatological data, which consists of 25 features, with a total of 76891 records, with one hour periodicity.

No missing values were detected in the datasets. An important point, as the study target only contains data from August 2018 to May 2020, then only data between these dates were considered. A monthly average of the total nitrogen variation was made for each year, illustrated in Fig. 1. Analyzing the figure, it is possible to verify that the highest total nitrogen value occurs mainly in the months corresponding to Spring and Summer, in 2018 and 2019. In addition, in all years in the Autumn and Winter months, the values tend to decrease. After

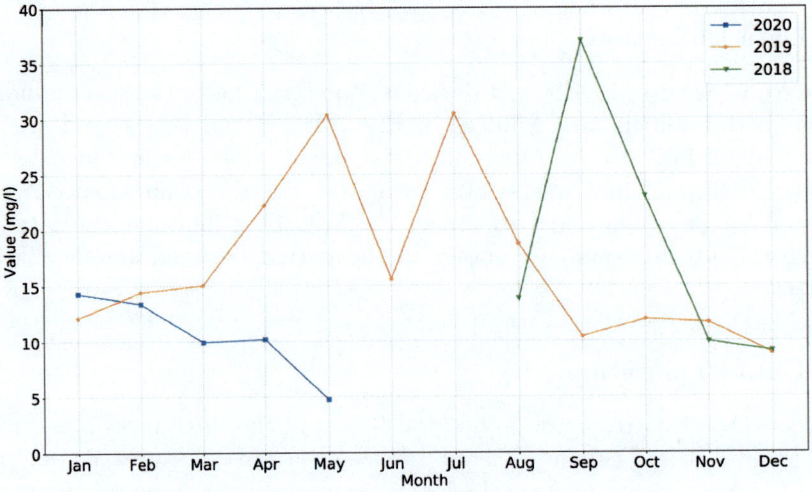

Fig. 1. Average variation in total nitrogen over the years.

that, the *Shapiro-Wilk* test was performed to verify if the features followed a normal distribution. With a $p < 0.05$, we concluded that all features assumed a non-Gaussian distribution.

3.3 Data Preparation

Instead of simply listing headings of different levels we recommend to let every heading be followed The first step in the data preparation process was to apply feature engineering to create three new features: *year*, *month* and *day*. After that, regarding placing the features with the same timesteps, it was necessary to use a kind of "one-hot encoding", where the indicators are represented by columns with the corresponding values on each date. For example, we created a column called "*total_nitrogen_affluent*" with the corresponding values for the registration of total nitrogen in the affluent. This change led to the appearance of missing values since some indicators present different periodicities and some records on different dates.

On average, the records of the indicators are over a period of 7 days. Therefore, all indicators were placed with a week average value, grouping them by week. With the same periodicity in all data, the missing values were replaced with the average values corresponding to the previous three weeks.

The next step consisted in the identification of the correlations between the features and target feature. Since all features didn't follow a normal distribution, we used the non-parametric Spearman's rank correlation coefficient. Some correlations were found with a target feature, such as between ammonia and target feature. Also, no sufficiently strong correlations were found between the target feature and the climatological data. Moreover, none of the features created through the feature engineering process have a strongly correlation with total nitrogen.

Lastly, we removed the features that didn't have a strong correlation with the target feature. At the end of this process, the final dataset was left with 218 records and six features. Table 2 describe an example of a record in the final dataset.

Table 2. Features present in the final dataset.

#	Features	Observation example
0	date	2016-01-04
1	total_nitrogen_affluent	31.638
2	cod_treated_effluent	12.515
3	tss_affluent	54.231
4	ammonia_treated_effluent	4.883
5	ostophosphates_treated_effluent	3.658
6	total_nitrogen_treated_effluent	7.708

3.4 Evaluation Metrics

To evaluate the conceived candidate models, we used two error metrics. One of them was RMSE, which is the standard deviation of the forecasting errors by the model. In other words, it's a measure of precision, where it measures the difference between the values predicted by the model and the real values. Its equation is as follows:

$$RMSE = \sqrt{\frac{\sum_{i=1}^{n}(y_i - \hat{y}_i)^2}{n}} \tag{1}$$

The MAE metric was also used, which averages the differences between the predicted values and the actual values. The use of this metric aims to reinforce the confidence of the values obtained through the model used. MAE equation is as follow:

$$MAE = \frac{1}{n}\sum_{i=1}^{n}|y_i - \hat{y}_i| \tag{2}$$

3.5 Decision Trees

The DT is a supervised learning algorithm in the form of a tree structure. It divides a dataset into smaller and smaller subsets, and, at the same time, an associated DT is developed incrementally. The result is a tree with decision and leaf nodes. A decision node has two or more branches, each representing values for the tested attribute. The leaf node represents a decision on the numerical or categorical target [13].

The main advantages of these models are the ease of interpretation of the results, being possible to visualize the main attributes for the problem, as they appear at the top of the tree. Also, DTs do not require data normalization and scaling, and work on both classification and regression problems. In contrast, a DT is very sensitive to small disturbances in the datasets [13].

3.6 Random Forests

A RF, like DTs, is a supervised learning algorithm. The "forest" it builds is a set of DTs, usually trained with the "bagging" method. The general idea of the bagging method is a combination of learning models that increases the overall result. The DTs created at RF are executed in parallel, with no interaction between these trees in the construction process. Also, a RF adds additional randomness to the model while "cultivating" the trees. Instead of looking for the most important feature when dividing the node, it looks for the best feature among a random subset of features. This results in a wide diversity that generally results in a more accurate model [14].

The main advantages of these models are the automatic treatment of lost values, as they are replaced by the variable that most appears in a given node. Although, RFs doesn't need feature scaling (standardization and normalization) and can deal with continuous and categorical variables. It can be used for classification and regression problems. However, they are more complex than DTs and, consequently, take a long time in the training process [14].

4 Experiments

To achieve the study's goal, it was necessary to conceive, tune, and evaluate several candidate models of DTs and RFs. For this, to find the best combination of hyperparameters for each model, several experiments were carried out. We used a GridSearchCV tool to search for the best hyperparameters of the models and make a cross-validation technique in each experiment, with a CV Split equal to 3. The value of both error metrics comes from the average of the obtained values in the three splits, with this value used to evaluate the performance of the candidate models.

The hyperparameters searched in each model have a great similarity, except for *n_estimators*, which is only used in RF, and the *splitter*, where only DT uses it. However, the values of the hyperparameters between the models are also different because the RF-based models are more complex than DT-based. Therefore, it may require higher values in hyperparameters to obtain better results. Table 3 describes the search for the hyperparameters considered for each of the models.

Regarding the programming language, we used *Python*, version 3.7, with some libraries, for all the steps carried out in this work.

Table 3. Hyperparameters searching space.

Parameter	DT	RF	Rationale
max_depth	[5,10]	[5,12]	Maximum depth
min_samples_split	[2,6]	[2,8]	Minimum samples required to split
min_samples_leaf	[2,4]	[2,6]	Minimum samples required to be at a leaf
max_features	[auto,sqrt,log2]	[auto,sqrt,log2]	Number of features for the best split
n_estimators	-	[20,100]	Number of trees in the forest
splitter	[best,random]	-	Strategy used to choose the split at each node

5 Results and Discussion

The obtained results are presented in Table 4, which depicts the top-3 results of the best candidate models for each ML model. The best candidate model was a DT-based model, with a RMSE of 2.207 and a MAE of 1.612. This model use a *max_depth* of 7, a *min_samples_leaf* of 2, and a *max_samples_split* of 3. Furthermore, the best RF candidate model obtained a RMSE of 2.271 and a MAE of 1.679. It is worth mentioning that the values of the hyperparameters are very homogeneous in the three best candidates models of DTs, such as *min_samples_leaf*. On the other hand, the bests RFs candidate models have a low level of homogeneity concerning the hyperparameters. It is also important to note that the RFs models use considerably more time in the training process when compared to the DTs models, which was already expected due to the greater complexity present in the RFs models.

Moreover, concerning the hyperparameter *max_depth* value, in the three best candidate DT-based models, we can verify this tends to increase with increasing error, while in the RF-based models this value tends to decrease with increasing error. Also, it is possible to verify, in the case of the *max_samples_split*, that it also tends to grow with the error increase in the DT-based models.

Table 4. DT and RF top-3 models. Legend: a - max_depth; b - max_samples_split; c - min_samples_leaf; d - max_features; e - n_estimators; f - splitter; g - RMSE; h - MAE; i - time (in seconds).

a	b	c	d	e	f	g	h	i
DT candidate models								
7	3	2	auto	-	best	**2.207**	**1.612**	**0.0135**
7	5	2	auto	-	best	2.855	2.116	0.0137
9	5	2	auto	-	best	2.869	2.163	0.0155
RF candidate models								
8	2	2	auto	50	-	**2.271**	**1.679**	**0.2271**
7	4	4	auto	80	-	2.915	2.266	0.1854
7	2	3	auto	60	-	2.983	2.362	0.2013

Figure 2 illustrates the forecast made by the best DT-based and RF-based candidate model. As can be seen, the predictions of both models are close, with superficial differences. Even so, it is clear that the DT-based model can obtain better forecasts than the RF model.

In addition to the forecast, the rules that represent the conditional statement were also verified from the best candidate model (DT-based). In Table 5, it is possible to check some of these rules, which describe some cases where the total nitrogen value exceed the maximum allowed by law, a value that cannot be greater than 15 mg/L [9]. It's possible to verify, for example, that the total nitrogen in the treated effluent reaches the value of 51.95 when the Biochemical Oxygen Demand (BOD) in the treated effluent is less than 47.06 and orthophosphates are less or equal to 0.76.

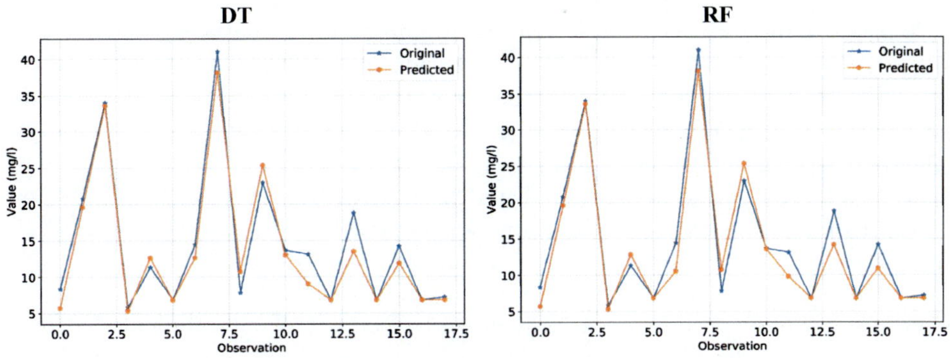

Fig. 2. Total nitrogen forecast using the best DT and RF models.

Table 5. Some rules defined by the best candidate model.

Rules	1	2	3	4	5
Total nitrogen afluent	-	≤85.35	≤85.35	-	-
BOD effluent	<47.06	<33.17	≤28.98	≤42.26	≤33.17
TSS afluent	-	≤326.56	≤326.56	-	<454.04
Ammonia effluent	-	-	<28.94	-	<11.25
Ostophosphates effluent	≤0.76	<0.76	-	<0.76	≤1.49
Total nitrogen effluent	**51.95**	**33.65**	**21.71**	**31.88**	**17.91**

6 Conclusions

In recent years, governments' campaigns for better nitrogen management have been increasing due to its consequences for the environment. Consequently, this work allows WWTP managers to check what is causing the unwanted values presented by the total nitrogen in the treated effluent and, thereby, prevent its consequences.

This study focused on forecast the total nitrogen in the treated effluent, using two ML models for that purpose. The rules that clarify the values foreseen by the best candidate model were also extracted. The model that obtained the best performance was a DT-based model with a MAE of 1.612 and a RMSE of 2.207, showing that it is possible to make accurate predictions. However, RF also obtained promising values, with the results being close to DT, revealing that it is also a valid model. The interesting point is that the simplest model achieves slightly better results than greater complexity on the RF part. This can be justified by the presence of a small dataset, where a DT can even surpass a RF, with significantly less processing time.

Therefore, future work will consider adding more input features, such as more indicators present in the WWTP. This inclusion may allow more correlations on the total nitrogen to be found. Another research to be done would be to apply the same approach, but instead of using total nitrogen as a target, use another indicator that has unwanted values to allow WWTP decision-makers to act as quickly and as objective as possible. Finally, we also intend to use Deep Learning models to forecast total nitrogen, namely, Recurrent Neural Networks (RNNs), which have shown good performance in time series problems.

Acknowledgements. This work is financed by National Funds through the Portuguese funding agency, FCT - Fundação para a Ciência e a Tecnologia within project DSAIPA/AI/0099/2019.

References

1. Fernandes, B., Vicente, H., Ribeiro, J., et al.: Fully informed vulnerable road users: simpler, maybe better. In Proceedings of the 21st International Conference on Information Integration and Web-based Applications & Services (iiWAS2019), pp. 598–602 (2019). https://doi.org/10.1145/3366030.3366089
2. Kunz, A., Peralta-Zamora, P., Moraes, S.G.D., Durán, N.: Novas tendências no tratamento de efluentes têxteis. Química nova **25**(1), 78–82. https://doi.org/10.1590/S0100-40422002000100014
3. Connor, R.: The United Nations World Water Development Report 2015: Water for Sustainable World (Vol. 1). UNESCO publishing (2015)
4. World Health Organization: Sanitation Safety Planning: Manual for Safe Use and Disposal of Wastewater Greywater and Excreta. World Health Organization (2015)
5. UN-Water, UNESCO: United Nations World Water Development Report 2020: Water and Climate Change (2020)
6. Oliveira, P., Fernandes, B., Analide, C., Novais, P.: Forecasting energy consumption of wastewater treatment plants with a transfer learning approach for sustainable cities. Electronics **10**, 1149 (2021). https://doi.org/10.3390/electronics10101149
7. Rutherford, P.M., McGill, W.B., Arocena, J.M., Figueiredo, C.T.: Total nitrogen. Soil Sampl. Methods Anal. **2**, 239–250 (2008)
8. Fernandes, B., Silva, F., Alaiz-Moretón, H., Novais, P., Neves, J., Analide, C.: Long short-term memory networks for traffic flow forecasting: exploring input variables. time frames and multi-step approaches. Informatica **31**(4), 723–749 (2020). https://doi.org/10.15388/20-INFOR431
9. Ministério do Ambiente.: Decreto-Lei n.o 152/97 (No. 152/97) (1997). https://data.dre.pt/eli/dec-lei/152/1997/06/19/p/dre/pt/html
10. Bagherzadeh, F., Mehrani, M.J., Basirifard, M., Roostaei, J.: Comparative study on total nitrogen prediction in wastewater treatment plant and effect of various feature selection methods on machine learning algorithms performance. J. Water Process Eng. **41**, 102033 (2021). https://doi.org/10.1016/j.jwpe.2021.102033
11. Guo, H., et al.: Prediction of effluent concentration in a wastewater treatment plant using machine learning models. J. Environ. Sci. **32**, 90–101 (2015). https://doi.org/10.1016/j.jes.2015.01.007
12. Nourani, V., Elkiran, G., Abba, S.I.: Wastewater treatment plant performance analysis using artificial intelligence-an ensemble approach. Water Sci. Technol. **78**(10), 2064–2076 (2018). https://doi.org/10.2166/wst.2018.477

13. Friedl, M.A., Brodley, C.E.: Decision tree classification of land cover from remotely sensed data. Rem. Sens. Environ. **61**(3), 399–409 (1997). https://doi.org/10.1016/S0034-4257(97)00049-7
14. Breiman, L.: Random forests. Mach. Learn. **45**(1), 5–32 (2001). https://doi.org/10.1023/A:1010933404324
15. Wood, A., Blackhurst, M., Hawkins, T., Xue, X., Ashbolt, N., Garland, J.: Cost-effectiveness of nitrogen mitigation by alternative household wastewater management technologies. J. Environ. Manage. **150**, 344–354 (2015). https://doi.org/10.1016/j.jenvman.2014.10.002

Check for updates

Machine Learning for Network-Based Intrusion Detection Systems: An Analysis of the CIDDS-001 Dataset

José Carneiro(✉), Nuno Oliveira, Norberto Sousa, Eva Maia, and Isabel Praça

Research Group on Intelligent Engineering and Computing for Advanced Innovation and Development (GECAD), Porto School of Engineering (ISEP), 4200-072 Porto, Portugal
{emcro,nunal,norbe,egm,icp}@isep.ipp.pt

Abstract. With the increasing amount of reliance on digital data and computer networks by corporations and the public in general, the occurrence of cyber attacks has become a great threat to the normal functioning of our society. Intrusion detection systems seek to address this threat by preemptively detecting attacks in real time while attempting to block them or minimizing their damage. These systems can function in many ways being some of them based on artificial intelligence methods. Datasets containing both normal network traffic and cyber attacks are used for training these algorithms so that they can learn the underlying patterns of network-based data. The CIDDS-001 is one of the most used datasets for network-based intrusion detection research. Regarding this dataset, in the majority of works published so far, the *Class* label was used for training machine learning algorithms. However, there is another label in the CIDDS-001, *AttackType*, that seems very promising for this purpose and remains considerably unexplored. This work seeks to make a comparison between two machine learning models, K-Nearest Neighbours and Random Forest, which were trained with both these labels in order to ascertain whether *AttackType* can produce reliable results in comparison with the *Class* label.

1 Introduction

In the last few years Intrusion Detection System (IDS) research has had a progressively increasing interest. The application of Artificial Intelligence (AI) methods for IDS has been widely considered in the literature [1] due to their ability to learn complex patterns inherent to network-based data. These patterns are later used in the deployment phase to timely detect probable attack attempts that threaten a network's normal functioning and privacy.

Reliable testbeds comprised of both normal network behaviour and attack scenarios are required to train AI algorithms and test their detection performance in controlled environments. The lack of well grounded datasets for the Network Intrusion Detection Systems (NIDS) setting has been appointed as one of the

© The Author(s), under exclusive license to Springer Nature Switzerland AG 2022
K. Matsui et al. (Eds.): DCAI 2021, LNNS 327, pp. 148–158, 2022.
https://doi.org/10.1007/978-3-030-86261-9_15

major drawbacks of current research [2]. However, in the last few years, some datasets have been proposed to solve this problem, namely the CTU 13 [3], the SANTA data set [4], the CICIDS-2017 [2], and the CIDDS-001 dataset [5].

IDS datasets can mainly be divided into two categories, packet-based and flow-based. The more conventional ones such as DARPA99 [6] or its improved version, KDD CUP 99 [7] are packet-based. These contain great amounts of information and features such as packet-headers and payloads [5]. Flow based datasets, however, such as CTU 13 [3] and CIDDS-001 [5] are usually more compact as their features are composed of aggregated information that was extracted from the communications within the network. These types of datasets were proposed by Wheelus and Zuech *et al.* [4] in 2014, as they are more recent and there are relativity fewer examples when compared to packet-based ones.

The CIDDS-001 is a recent flow-based dataset for the IDS setting. It contains unidirectional *Netflow* data and was generated using Python scripts to simulate human behaviour on virtual machines of an emulated network. It is very realistic since it respects operating cycles and working hours in enterprises. It contains both normal data and different types of cyber attacks, namely ping scans, port scans, brute forces and denial of services (DoS). Since the technologies used to generate the attacks are time-dependent the flows of the dataset were labeled based on their timestamp. Four different labels were considered, *Class*, which classifies the flow into normal, attacker, victim, suspicious or unknown, *AttackType*, which represents the type of the executed attack, *AttackID*, which contains the ID of the attack instance and *AttackDescription*, which provides a short description with attack-related details.

In this work, a comparison between the performance of two Machine Learning (ML) models, Random Forest (RF) and K-Nearest Neighbors (KNN), was performed under the CIDDS-001 setting. These models were chosen to conduct this study because they are widely used in the NIDS setting exhibiting great performances in several testbeds [8,9].

Most studies around this dataset have used the *Class* label as target variable, with one found exception [10], which used *AttackType*. Therefore, this study seeks to compare both labels and to evaluate if the *AttackType* is a reliable target to train and evaluate ML algorithms in the CIDDS-001 context.

For a SOC operator, knowing that an attack is occurring is extremely important, but knowing the type of attack is equally important. This knowledge will influence his following decisions as well as what measures to take in order to mitigate the threat. Therefore, a system that is capable of not only detecting attacks but also classifying them is of great use for any IDS.

This paper is organized into multiple sections. Section 2 presents an overview of the current state of the literature regarding NIDS research on the CIDDS-001 dataset. Section 3 describes the algorithms and testbed chosen to perform this study. Section 4 presents the achieved results and their discussion. Section 5 provides the main conclusions of this work as well as future research topics to be consequently addressed.

2 Related Work

Several anomaly detection approaches based on ML have been proposed in the context of NIDS. Many works have selected the CIDDS-001 as testbed to train and test their algorithms. Most of them used the *Class* label as target variable, presenting outstanding results in model performance. To the best of our knowledge, only one work, [10], has used the *AttackLabel* label.

In [11], Verma *et al.* performed a statistical analysis of the CIDDS-001 dataset using KNN and K-means clustering. These models were trained using the *Class* label, classifying each flow as either suspicious, attack, victim, unknown or normal. The data used for model training was separated into data from the *External server* and data from the *OpenStack*. The algorithms trained on both sets achieved extremely good results with an accuracy value close to 100%.

In [12], Althubiti *et al.* trained a Long-Short Term Memory (LSTM) using the CIDDS-001 dataset and the *Class* label, achieving an accuracy of almost 85% and a recall and precision of almost 86% and 88%, respectively. They compared this model with a Support Vector Machine (SVM), a Naïve Bayes and a Multi-Layer Perceptron (MLP) which achieved slightly worse results than the LSTM, around 80%.

In [13], Verma *et al.* tested a variety of ML models to detect DoS attacks in the context of internet of things. The CIDDS-001, UNSW-NB15, and NSL-KDD were used for training and benchmarking with a wide variety of models such as RF, AdaBoost, gradient boosted machine, regression trees and MLPs. The models achieved a good performance in detecting DoS attacks for the CIDDS-001 dataset with the RF presenting an accuracy of almost 95% and an Area Under the Curve (AUC) of almost 99%.

In [14], I. Kilincer *et al.* used five of the most widely acknowledged IDS datasets, CSE-CIC IDS-2018, UNSW-NB15, ISCX-2012, NSL-KDD and CIDDS-001, and trained a SVM, a KNN and a Decision Tree with each of them. Great results were achieved for every algorithm, with the KNN trained with CIDDS-001 dataset achieving an accuracy, recall, precision and f1-score of approximately 97%. The model that achieved the best results with this dataset was the Decision Tree with all four referred metrics above 99%.

In [15] Zwane *et al.* performed an analysis of an ensemble learning approach for a flow based IDS using the CIDDS-001 dataset. A variety of ensemble learning approaches were employed. These techniques were applied on three algorithms, Decision Tree, Naïve Bayes and SVMs. A RF was also implemented. Results greatly differed based on the chosen algorithm and ensemble technique. The best performing algorithms were the RF and all the ensembles of decision trees with an accuracy, precision, recall and f1-score of 99%. The ensembles trained with the Naïve Bayes and the SVMs performed worse, with all four metrics varying between 60% and 70%.

In [16], Tama and Rhee performed an attack classification analysis of IoT network by using deep neural networks for classifying attacks trained using the CIDDS-001, UNSW-NB15 and GPRS datasets. The deep neural network trained with the CIDDS-001 achieved excelent results, with an accuracy, precision and recall of 99%, 99% and 100% respectively.

In [10], N. Oliveira *et al.* proposed a sequential approach to intrusion detection by using time-based transformations in the algorithms' input data. The CIDDS-001 dataset was used for training three distinct algorithms, LSTM, RF and MLP. Additionally, the *AttackType* label was chosen instead of the *Class* label. The introduced approach achieved 99% accuracy for both the LSTM and the RF. An F1-score of near 92% was obtained for the LSTM.

As far as we are aware, this is the first work to perform a comparison between both labels in an attempt to determine if any gain can be obtained from using *AttackType* instead of the *Class* label.

3 Materials and Methods

This section presents the CIDDS-001 dataset and the way it was labelled, as described by it's authors in the technical report [17]. The employed algorithms, RF and KNN, parameters and configurations are described as well as the evaluation metrics chosen for validation and comparison.

3.1 Dataset Description

The CIDDS-001 (Coburg Intrusion Detection Data Set) was developed by Markus Ring *et al.* [5] in 2017. This dataset is flow-based unlike most conventional datasets, which are packet-based.

The data present in the CIDDS-001 dataset can mainly be divided into two sets, based on how it was generated. Data generated through *OpenStack* emulating a small business environment and data generated through *External Server*, capturing real network data. The dataset contains a total of 31959175 flows, with 31287934 being generated by the *OpenStack* and 671241 from the *External server*. A total of 33 million flows comprises the CIDDS-001, spanning a total of four weeks of network traffic, from March 3 2017 to April 18 2017. All 33 millions flows are labelled, containing benign network traffic as well as four different types of attacks, Ping Scans, Port Scans, DoS and Brute force.

Unlike packet-based datasets, which usually contain a lot of features, the CIDDS-001 contains a total of 12 with an additional 4 labels. These are standard *Netflow* features, namely source and destination IPs and ports, transport protocol, duration, date first seen, number of bytes, number of packets, number of flows, type of service and TCP flags. The dataset contains an additional four labels, *Class*, *AttackID*, *AttackType* and *AttackDescription*.

3.2 Dataset Labelling

The labelling process of the CIDDS-001 is composed of two steps, due to the different data sources. Both the traffic from the *External Server* and the one generated by the *OpenStack* environment were processed and labelled accordingly. Since the *OpenStack* traffic was generated in a fully controlled network this involved using the timestamps, origins and targets of the executed attacks [17], marking the remaining traffic as normal. A representation of the simulated small business environment is described in Fig. 1.

As for the traffic originated from the *External Server*, its labelling process was not as simple. All traffic incoming into this server from the *OpenStack* environment was benign, and as such was labelled as normal. The additional traffic incoming from the three controlled servers, *attacker1, attacker2 and attacker3*, is only malicious, so the traffic incoming and outgoing from these servers is labelled as attacker or victim respectively. All traffic from ports 80 and 443 is labelled as unknown since it comes from a homepage available to anyone interested in visiting. And finally all remaining traffic in the *External server* is labelled as suspicious [17].

The latter labels, unknown and suspicious, due to their uncertain nature, pose a problem when training the models. Considering that this traffic can include both normal and attack entries, and that there is no way to correctly categorize each one, the use of the whole environment as either normal or attack will create a strong bias towards these classes.

3.3 Dataset Preprocessing and Sampling

In preparation for the ML methods training, the CIDDS-001 has to go through a series of preprocessing procedures. This process was largely based on the work of N. Oliveira *et al.* [10] that implemented a preprocessing pipeline consisting of several steps. A visual representation of the overall process is presented in Fig. 2. An initial analysis to detect errors and inconsistent data was made, eliminating features such as *Flows* that contains a single value throughout the whole dataset and correcting others such as *Bytes* where suffixes like K and M were replaced by their numerical representation, in order to preserve the sequence of the flow, the dataset was indexed with the *Date first seen* feature. Non-numerical features were encoded using categorical encoding. After applying this pipeline, there were 10 features remaining, *Src IP*, *Src Port*, *Dest IP*, *Dest Port*, *Proto*, *Flags*, *ToS*, *Duration*, *Bytes* and *Packets*, all of them encoded as numbers and scaled between 0 and 1. The sampling process was likewise performed in the same manner as N. Oliveira *et al.* [10], resulting in a representative sample of the whole dataset. Since the resulting sample is a lot less demanding in terms of memory, processing power and time but still very representative of the dataset as a whole. The sample contains a total of 2,535,456 flows between 2:18:05 p.m., 17 March 2017and 5:42:17 p.m., 20 March 2017. This final sample was then split into a training and testing set using a simple holdout method where 70% of the data was used for training and 30% for testing.

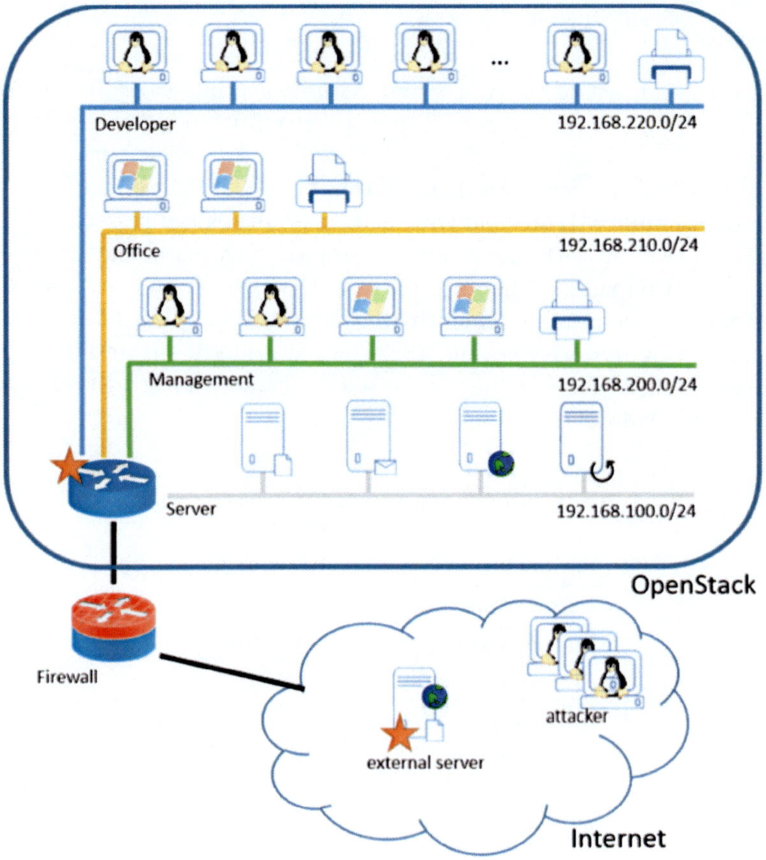

Fig. 1. Schematic of the emulated small business environment [17]

Since this work seeks to compare algorithms trained with both the *Atttack-Type* and *Class* labels, an additional preprocessing step was performed for the latter. All instances of suspicious, unknown and victim were replaced with the attack class, transforming the problem as a binary classification task with only two categories, normal and attack.

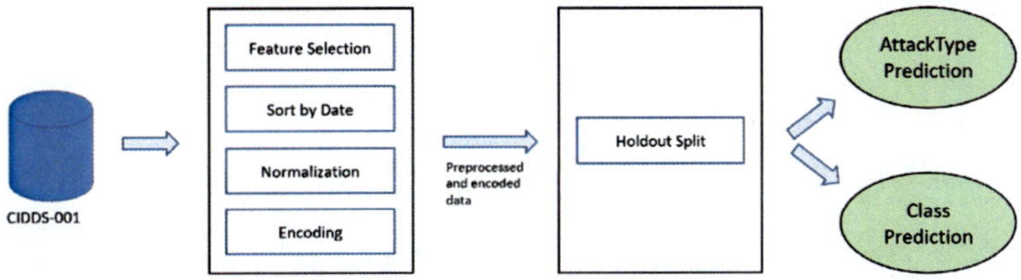

Fig. 2. Methodology workflow

3.4 Models

To compare the impact of the *Class* and *AttackType* labels in the ML methods performance, two algorithms were selected to be employed in this work, RF and KNN. These models have been thoroughly studied by the scientific community presenting good results for several IDS datasets [18].

RF is an ensemble ML method that makes use of several instances of decision trees which are trained with different sub-samples of the same dataset to produce diverse classification rules. When presented with a sample to be classified, all the decision trees of the ensemble make their own prediction. The final classification result is decided by majority voting [19]. The parameters selected for the RF are described in Table 1. These values were selected based on grid search within a range of possible values.

Table 1. Random forest classifier parameters.

Parameter	Value
Nº of estimators	10
Split criterion	Gini impurity
Min samples split	2
Min samples leaf	1
Max features	$\sqrt{n_{features}}$
Min impurity decrease	0
Class weight	Balanced

On the other hand, KNN is a supervised ML algorithm used for both classification and regression. When used as a classifier, the algorithms prediction is the highest frequency class among the sample's k nearest data points. The distance between the dataset instances is calculated using a specific metric such as Euclidean and Manhattan. The parameters selected for the KNN were also obtained by performing a grid search optimization and are presented in Table 2.

Table 2. KNN classifier parameters.

Parameter	Value
Nº of neighbors	3
Weights	Uniform
Leaf size	30
Metric	Minkowski

4 Results and Discussion

Generally a robust evaluation of ML models considers several metrics [20]. In this work four metrics were used: accuracy, precision, recall and F1-score [21]. Accuracy is a generalist metric that usually gives a good indication of an algorithm's performance. Precision measures the amount of positive predictions that were actually negative. Recall determines how many of the total positive label instances were correctly labelled. F1-score is calculated using both recall and precision, being a good metric to be accounted when a balance between these metrics is intended. Additionally, this metric is extremely reliable when dealing with unbalanced datasets, such as CIDDS-001, since it is not biased towards the majority class [20].

To perform the required computations, Google Colab [22] was used as hardware support providing free access to considerable amounts of disk space and RAM. However, the hardware is not unlimited and there are no guarantees that it remains the same over time. Therefore, a comparison between the models computing time cost could not be performed.

4.1 Label Comparison

The results obtained for the *Class* label were near 100% for all metrics for both models, RF and KNN. On the other hand, the ones regarding the *AttackType* label are quite distinct being highly influenced by the algorithms performance on the minority classes, such as ping scan and brute force. The macro-averaged results are described in Table 3.

Table 3. AttackType label macro average model results.

Model	Accuracy	Precision	Recall	F1-score
Random Forest	0.9560	0.9032	0.9239	0.9134
KNN	0.9694	0.9037	0.9290	0.9161

Table 4 shows the results of the trained RF model for the classification of each type of attack with the *AttackType* label.

Table 4. AttackType label random forest model results.

Class	Precision	Recall	F1-score
Brute force	0.9791	0.9868	0.9239
DoS	1.0000	0.9999	1.0000
No attack	0.9999	1.0000	1.0000
Ping scan	0.8104	0.5344	0.6441
Port scan	0.9906	0.9950	0.9928

Table 5 displays the same information but for the trained KNN model.

Table 5. AttackType label KNN model results.

Class	Precision	Recall	F1-score
Brute force	0.9920	0.9815	0.9867
DoS	0.9999	0.9999	0.9999
No attack	0.9999	1.0000	0.9999
Ping scan	0.8650	0.5406	0.6654
Port scan	0.9900	0.9965	0.9932

4.2 Discussion

As stated in [17], the labelling process of the traffic incoming from the *External Server* in the CIDDS-01 dataset with regards to the *Class* label isn't accurate, as all traffic incoming from ports 80 and 443 is labelled as either unknown or suspicious without any real certainty of whether that flow relates to an attack or not. Additionally a score of nearly 100% for all metrics in both models trained with this label is abnormal and not coherent, therefore it is highly probable that these models are overfitted and that these results do not reflect their intrusion detection capability.

As for the results of the *AttackType* label, they appear to be very promising since the labelling process accurately identifies all the attacks that were performed in the testbed. This presents a great advantage over the *Class* label since it assures a more robust and less biased classifier. The obtained results are also lower in terms of macro F1-score since it is typically hard for machine learning algorithms to perform well for the minority classes of unbalanced datasets such as CIDDS-001. Nevertheless, both models performed quite well for all attack types with the exception of the ping scan class.

5 Conclusion

This work has established a comparison between two labels, *Class* and *Attack-Type*, of the CIDDS-001 dataset, a widely regarded testbed for NIDS research. Two ML algorithms, RF and KNN, were trained with both these labels in order to compare their performance as classifiers and assure that the *AttackType* can be a reliable target variable although *Class* is more commonly used in previous works.

The results for the *Class* label were near 100% for all metrics for both models, seemingly too perfect, suggesting the occurrence of over-fitting. Differently, although the results regarding *AttackType* are slightly lower in terms of absolute values, these seem to be a lot more promising since the labeling process assures the correct identification of all attacks performed in the testbed. The KNN achieved the best F1-score value, 91.61%, slightly above the one presented by the RF, 91.34%.

This research, to the best of our knowledge, is the first to address a comparison between both these labels and to appoint that the *Class* labelling process is quite less reliable than the one for the *AttackType*. As future work, the *AttackType* label will be explored in greater detail by experimenting with other ML algorithms in an attempt to improve the presented results.

Acknowledgements. This work has received funding from European Union's H2020 research and innovation programme under SAFECARE Project, grant agreement no.787002.

References

1. Sultana, N., Chilamkurti, N., Peng, W., Alhadad, R.: Survey on SDN based network intrusion detection system using machine learning approaches. Peer-to-Peer Network. Appl. **12**(2), 493–501 (2018). https://doi.org/10.1007/s12083-017-0630-0
2. Sharafaldin, I., Habibi Lashkari, A., Ghorbani, A.A.: toward generating a new intrusion detection dataset and intrusion traffic characterization. In: Proceedings of the 4th International Conference on Information Systems Security and Privacy, pp. 108–116. SCITEPRESS - Science and Technology Publications (2018)
3. García, S., Grill, M., Stiborek, J., Zunino, A.: An empirical comparison of botnet detection methods — Elsevier Enhanced Reader
4. Wheelus, C., Khoshgoftaar, T.M., Zuech, R., Najafabadi, M.M.: a session based approach for aggregating network traffic data - the SANTA dataset. In: 2014 IEEE International Conference on Bioinformatics and Bioengineering, pp. 369–378, November 2014
5. Ring, M., Wunderlich, S., Grüdl, D., Landes, D., Hotho, A.: Flow-based benchmark data sets for intrusion detection. In: Proceedings of the 16th European Conference on Cyber Warfare and Security (ECCWS), p. 10 (2017)
6. Thomas, C., Sharma, V., Balakrishnan, N.: Usefulness of DARPA dataset for intrusion detection system evaluation. In: Proceedings of SPIE - The International Society for Optical Engineering (2008)
7. Tavallaee, M., Bagheri, E., Lu, W., Ghorbani, A.: A detailed analysis of the KDD CUP 99 data set. In: IEEE Symposium. Computational Intelligence for Security and Defense Applications, CISDA, vol. 2, July 2009
8. Maia, E., et al.: Cyber threat monitoring systems - comparing attack detection performance of ensemble algorithms. In: Abie, H., et al. (eds.) CPS4CIP 2020. LNCS, vol. 12618, pp. 31–47. Springer, Cham (2021). https://doi.org/10.1007/978-3-030-69781-5_3
9. Kumar, I., Mohd, N., Bhatt, C., Sharma, S.: Development of IDS using supervised machine learning. In: Pant, M., Kumar Sharma, T., Arya, R., Sahana, B., Zolfagharinia, H. (eds.) Using Sub-sequence Information with kNN for Classification of Sequential Data, pp. 565–577. Springer, Singapore (2020). https://doi.org/10.1007/978-981-15-4032-5_52

10. Oliveira, N., Praça, I., Maia, E., Sousa, O.: Intelligent cyber attack detection and classification for network-based intrusion detection systems. Appl. Sci. **11**(4), 1674 (2021)
11. Verma, A., Ranga, V.: Statistical analysis of CIDDS-001 dataset for network intrusion detection systems using distance-based machine learning. Procedia Comput. Sci. **125**, 709–716 (2018)
12. Althubiti, S.A., Jones, E.M., Roy, K.: LSTM for anomaly-based network intrusion detection. In: 2018 28th International Telecommunication Networks and Applications Conference (ITNAC), pp. 1–3 (2018)
13. Verma, A., Ranga, V.: Machine learning based intrusion detection systems for IoT applications. Wirel. Pers. Commun. **111**(4), 2287–2310 (2019). https://doi.org/10.1007/s11277-019-06986-8
14. Kilincer, I.F., Ertam, F., Sengur, A.: Machine learning methods for cyber security intrusion detection: datasets and comparative study. Comput. Netw. **188**, 107840 (2021)
15. Zwane, S., Tarwireyi, P., Adigun, M.: Ensemble learning approach for flow-based intrusion detection system. In: 2019 IEEE AFRICON, pp. 1–8 (2019)
16. Adhi Tama, B., Rhee, K.H.: Attack classification analysis of IoT network via deep learning approach. Res. Briefs Inf. Commun. Technol. Evolu. (ReBICTE), vol. 3, November 2017
17. Ring, M., Wunderlich, S., Grudl, D.: Technical report CIDDS-001 data set. J. Inf. Warfare 13 (2017)
18. Anbar, M., Abdullah, R., Hasbullah, I.H., Chong, Y.-W., Elejla, O.E.: Comparative performance analysis of classification algorithms for intrusion detection system. In: 2016 14th Annual Conference on Privacy, Security and Trust (PST), Auckland, New Zealand, pp. 282–288, IEEE, December 2016
19. Biau, G., Scornet, E.: A random forest guided tour. TEST **25**(2), 197–227 (2016). https://doi.org/10.1007/s11749-016-0481-7
20. Handelman, G.S., et al.: Peering into the black box of artificial intelligence: evaluation metrics of machine learning methods. Am. J. Roentgenol. **212**, 38–43 (2019)
21. Hossin, M., Sulaiman, M.N.: A review on evaluation metrics for data classification evaluations. Int. J. Data Min. Knowl. Manage. Process 5, 1–11 (2015)
22. Bisong, E.: Google Colaboratory, pp. 59–64. Apress, Berkeley (2019)

Wind Speed Forecasting Using Feed-Forward Artificial Neural Network

Eduardo Praun Machado[1,4], Hugo Morais[2], and Tiago Pinto[3](✉)

[1] Instituto Superior Técnico-IST, Universidade de Lisboa, 1049-001 Lisbon, Portugal
[2] Department of Electrical and Computer Engineering, INESC-ID, Instituto Superior Técnico-IST, Universidade de Lisboa, 1049-001 Lisbon, Portugal
hugo.morais@tecnico.ulisboa.pt
[3] GECAD – Research Group on Intelligent Engineering and Computing for Advanced Innovation and Development, ISEP/IPP, Polytechnic of Porto, Porto, Portugal
tcp@isep.ipp.pt
[4] Centro de Pesquisa de Energia Elétrica (Cepel), Cidade Universitária, Ilha do Fundão, Avenida Horácio Macedo, 354, Rio de Janeiro, RJ 21941-911, Brasil

Abstract. This paper presents a novel feed-forward neural network for wind speed forecasting. The electricity sector accounts for a quarter of the world CO_2 emissions. To reduce these emissions, several national, regional and global agreements have been signed, setting ambitious goals to increase the penetration of renewable energy sources (RES). Although achieving those goals is essential for the sector decarbonization and, therefore, to mitigate the global climate crisis, renewable-based generation can depend on highly variable and uncertain resources, such as the wind. Hence, having access to reliable forecasts of those resources availability is essential for the operation of several actors in the power and energy sector, and for the effectiveness of the whole system. This paper contributes to surpass this problem by introducing a new forecasting model based on a feed-forward neural network to forecast wind speed. The proposed model is applied to real data from a wind farm in the south of South America. Results show that the proposed model can achieve lower forecasting errors than the baseline models, which consist of Numerical Weather Predictions.

Keywords: Feed-forward neural network · Forecasting · Machine learning · Power and energy systems · Wind speed

1 Introduction

Over the last decades, power production from renewable sources have experienced a sharp expansion. With growing concerns about the global climate crisis, the expectation is that this trend will continue to escalate. Amid those sources, wind energy arises as one of the most attractive due to its high generation capacity, efficiency and cost-benefit ratio. However, as others RES, wind power generation also suffers from resource volatility and intermittency which imposes a challenge to its large-scale penetration as it can undermine the whole electrical system operation [1].

© The Author(s), under exclusive license to Springer Nature Switzerland AG 2022
K. Matsui et al. (Eds.): DCAI 2021, LNNS 327, pp. 159–168, 2022.
https://doi.org/10.1007/978-3-030-86261-9_16

To surpass these issues, accurate wind speed forecasts can play an essential role. They can, for example, help to optimize the market prices, producer profits [2] and electricity supply reliability [3]. Several factors, such as the environmental conditions, weather and time of the day can affect the predictions [4]. Nonetheless, the random and unstable characteristics of the wind hamper forecasts models to be precise [5]. Hence, minimizing the uncertainty associated with the wind is a keystone to improve wind energy forecasts [4].

This paper presents a wind speed forecasting model based on a Feed-forward Neural Network (FFNN), which can identify and learn patterns in wind speed variation. The proposed model was applied to the forecasting of the wind speed from a real wind turbine in a wind farm in the south of South America. To do this, it uses historical data from the respective Supervisory Control and Data Acquisition (SCADA) system and from the European Centre for Medium-Range Weather Forecasts (ECMWF) [6]. The results showed that the forecasts achieved by the proposed model reached lower errors than the baseline models, which are commonly employed by wind farm operators.

After this introductory section, Sect. 2 presents an overview of related work in the scope of wind speed forecasting. Section 3 presents the proposed FFNN model and Sect. 4 describes the used database. Section 5 presents some of the achieved results, and finally Sect. 6 wraps up the paper with the most relevant conclusions and contributions of this work.

2 Related Works

In [7], the authors proposed a bidirectional gated recurrent unit neural network (GRUNN) based model for NWP wind speed error correction and used the adjusted wind speed to forecast wind power, up to 24 h in advance. The model used NWP wind speed error standard deviation as a weight time series, which was later decomposed into trend and detail terms by means of Empirical Mode Decomposition. These two terms and the NWP wind speed were taken as inputs for the GRUNN so that corrections in the latter can be made. Finally, the corrected forecast was used to estimate wind power using the wind turbine wind power curve.

In [8], an NWP downscaling model with two different configurations were tested for hourly and sub-hourly 100-m wind speed forecasts: one using variables that described the boundary layer, winds and temperature which were available from the NWP outputs and another adding the error between measured and NWP wind speed as an input. The downscaling model uses a parametric approach with linear regression, which was developed in [9], coupled with stepwise regression based on Bayesian Inference Criterion. To verify the methodology, two years data from a wind farm close to Paris, France, was used and the results were compared with the original NWP forecast and benchmark models, namely Auto Regressive Moving Average, Artificial Neural Networks (ANN) and persistence.

In [10], a short-term wind power forecasting with NWP adjustment model was proposed. The authors developed a framework composed of three modules: wind power forecast, abnormality detection and data adjustment. Results show that the proposed model was able to correctly identify abnormal forecasts and to reduce the wind power forecast root mean squared error (RMSE) compered to the same method without the adjustment step and with the persistence model. It is noticed that the forecast error increased for longer horizons, which is also a finding in [8].

In [11], day-ahead wind power forecasts were made in a two-stage approach which was based on the combination of Hilbert-Huang Transform, Genetic Algorithm and ANN. Initially, the first ANN used NWP meteorological variables as inputs to predict the wind speed at hub height recorded by the SCADA system. In the second stage, another ANN was trained to map the wind power curve characteristics using historical SCADA records and uses the first stage forecasts to predict the wind farm output power. The model performed better than four other approaches in the four seasons of the year. The impact of input data on forecasts, namely wind speed, wind direction, air temperature, air pressure and air humidity is analyzed and an improvement is verified when a new variable is included.

Although these works present relevant findings and contributions in the domain of wind speed forecasting, the high variability of this type of resource and the increasing need for reliable forecasts from multiple entities in the power and energy sector, there is still a demand for further development to reach models with lower forecasting errors. This paper contributes to such error reduction, by proposing a novel model for wind speed forecasting using a FFNN.

3 Feed-Forward Artificial Neural Network

The considered FFNN is composed of nodes (or neurons) that are distributed across different layers, namely input, hidden and output layers. Each node in a layer is linked to the ones in the next by means of a weight parameter that measures the strength of that connection, forming a fully connected network structure that reminds the nervous system. The operating principle of neural networks can be described as a sequence of functional transformations [12]. For a given layer $l \in \{1, \ldots, L\}$, where L is the number of layers, a quantity called the activation value of the next layer $A^{[l]} = \left(a_1^{[l]}, \ldots, a_j^{[l]}\right)$ can be calculated as a linear combination of inputs $X^{[l-1]} = \left(x_1^{[l-1]}, \ldots, x_i^{[l-1]}\right)^T$ and weights $W^{[l]}$ in the form

$$A^{[l]} = W^{[l]} X^{[l-1]} + b^{[l]} \tag{1}$$

where

$$W^{[l]} = \begin{pmatrix} w_{1,1}^{[l]} & w_{1,2}^{[l]} & \cdots & w_{1,j}^{[l]} \\ w_{2,1}^{[l]} & w_{2,2}^{[l]} & \cdots & w_{2,j}^{[l]} \\ \vdots & \vdots & \ddots & \vdots \\ w_{i,1}^{[l]} & w_{i,2}^{[l]} & \cdots & w_{i,j}^{[l]} \end{pmatrix} \tag{2}$$

and $b^{[l]} = \left(b_1^{[l]}, \ldots, b_D^{[l]}\right)^T$ is a parameter called bias, which is used to adjust the output. The subscripts i and j represents the number of nodes or dimension of layers $l - 1$ and l, respectively.

The activation value $A^{[l]}$ is transformed by a nonlinear, differentiable function $h^{[l]}$ that is named activation function as in Eq. (3), resulting in the next layer input vector $X^{[l]}$. For hidden layers, the activation function is a logistic sigmoid or hyperbolic tangent function, while for the output layer it is the identity function.

$$X^{[l]} = h^{[l]}\left(A^{[l]}\right) \tag{3}$$

Equations (1) and (3) present recursive calculations that constitutes a process known as forward propagation [13]. This name comes from the fact that the information is flowing forward through the network, and this is the reason why this type of model is called FFNN. There are some particularities about these equations that should be mentioned: the first input vector $X^{[0]}$ comes from the features selected from the dataset, while the remaining are the result of the calculations. Also, the result observed in $X^{[L]}$ is the output of the model.

The parameters optimization is done with gradient descent-based calculations. With this approach, the required partial derivatives are related to the two parameters of a FFNN model: the weights and biases. Appling Eq. (4) to the last layer of this model results in (6) and (7).

$$p^{[i+1]} = p^{[i]} - \eta \nabla E(p) \tag{4}$$

where p represents the model parameters, η is the learning rate, and $E(p)$ is the loss function (5), which is the Mean Squared Error, also called Euclidean or L2 norm. In this equation, x_i is the forecasted value, t_i is the target value and n is the number of points in the dataset. The search for the loss function minimum is commonly done by computing its gradient ($\nabla E(p)$), which is the vector containing the partial derivatives of $E(p)$ [14]. The partial derivative $\frac{\partial}{\partial p}E(p)$ indicates how the function changes with a small change in one of the parameters. Therefore, the gradient vector points to the direction of the steepest increase of the function. As the learning algorithm goal is to minimize the error, with this approach the parameters can be updated at each iteration i by going in the opposite direction.

$$E(p) = \text{MSE} = \frac{1}{n}\sum_{i=1}^{n}(f(x_i, p) - t_i)^2 \tag{5}$$

$$W^{[L]}_{[i+1]} = W^{[L]}_{[i]} - \eta \frac{\partial E}{\partial W^{[L]}} \tag{6}$$

$$b^{[L]}_{[i+1]} = b^{[L]}_{[i]} - \eta \frac{\partial E}{\partial b^{[L]}} \tag{7}$$

Using the chain rule, one can write these partial derivatives as (8), (9)

$$\frac{\partial E}{\partial W^{[L]}} = \frac{\partial E}{\partial X^{[L]}} \circ \frac{\partial X^{[L]}}{\partial A^{[L]}} \cdot \frac{\partial A^{[L]}}{\partial W^{[L]}}^{T} \tag{8}$$

$$\frac{\partial E}{\partial b^{[L]}} = \frac{\partial E}{\partial X^{[L]}} \circ \frac{\partial X^{[L]}}{\partial A^{[L]}} \cdot \frac{\partial A^{[L]}}{\partial b^{[L]}}^{T} \tag{9}$$

where the dot (.) symbol stands for matrix multiplication and the circle (∘) symbol for the Hadamard or element-wise product. At this stage, it is useful to introduce the following notation (10)

$$\delta^{[L]} = \frac{\partial E}{\partial X^{[L]}} \circ \frac{\partial X^{[L]}}{\partial A^{[L]}} \tag{10}$$

where $\delta^{[L]}$ is a value known as delta and represents the error that the layer $L - 1$ sees.

Moving on to the layer $L - 1$, the calculations are as in (11), (12), (13).

$$\frac{\partial E}{\partial W^{[L-1]}} = \frac{\partial A^{[L]}}{\partial X^{[L-1]}}^{T} \cdot \left(\frac{\partial E}{\partial X^{[L]}} \circ \frac{\partial X^{[L]}}{\partial A^{[L]}} \right) \circ \frac{\partial X^{[L-1]}}{\partial A^{[L-1]}} \cdot \frac{\partial A^{[L-1]}}{\partial W^{[L-1]}}^{T} \tag{11}$$

$$\frac{\partial E}{\partial b^{[L-1]}} = \frac{\partial A^{[L]}}{\partial X^{[L-1]}}^{T} \cdot \left(\frac{\partial E}{\partial X^{[L]}} \circ \frac{\partial X^{[L]}}{\partial A^{[L]}} \right) \circ \frac{\partial X^{[L-1]}}{\partial A^{[L-1]}} \cdot \frac{\partial A^{[L-1]}}{\partial b^{[L-1]}}^{T} \tag{12}$$

$$\delta^{[L-1]} = \frac{\partial A^{[L]}}{\partial X^{[L-1]}}^{T} \cdot \delta^{[L]} \circ \frac{\partial X^{[L-1]}}{\partial A^{[L-1]}} \tag{13}$$

From layer $L - 1$ to the first layer, it is possible to write the next deltas as (14)

$$\delta^{[l]} = \frac{\partial A^{[l+1]}}{\partial X^{[l]}}^{T} \cdot \delta^{[l+1]} \circ \frac{\partial X^{[l]}}{\partial A^{[l]}} \tag{14}$$

and, therefore, the partial derivatives can be obtained as (15), (16)

$$\frac{\partial E}{\partial W^{[l]}} = \delta^{[l]} \cdot \frac{\partial A^{[l]}}{\partial W^{[l]}}^{T} \tag{15}$$

$$\frac{\partial E}{\partial b^{[l]}} = \delta^{[l]} \cdot \frac{\partial A^{[l]}}{\partial b^{[l]}}^{T} \tag{16}$$

Finally, the parameters updates can be performed using Eqs. (15) and (16) with (6) and (7) respectively. The process presented above constitutes the backpropagation learning algorithm. With this approach, the algorithm goes through each layer in reverse, measuring the error contribution from each connection by means of the deltas and updating the parameters accordingly [15]. By computing the gradient in reverse, the backpropagation algorithm avoids unnecessary calculations as it reuses previous ones. This is the major reason of this method higher computational effectiveness when compared to numerical methods such as finite differences [12] and one of the cornerstones to FFNN popularity.

4 Database

The database consists of NWP data from the ECMWF [6] and historical data from the SCADA system of a wind turbine in a wind farm in the south of South America. The NWP data collected in this study was the U and V components of wind, which are the components parallel to the x and y-axis respectively, at the wind farm location and at pressure level 133 (geopotential and geometric altitude of 106.54 m), as this height was the closest to the hub height.

The SCADA is a system that connects software and hardware elements of a wind turbine/farm in a single control system which collects, process and displays all measured data, allowing the operator to monitor turbine/farm conditions in real time. These data come from sensors and controllers installed in every subassembly of a wind turbine and it is usually sampled every 10-min, providing the average, maximum, minimum and standard deviation of the period. Among the recorded parameters are production status, such as active and reactive power; electrical data, such as voltages and currents; generator data, such as generator speed and bearing temperature; and environment data, such as wind speed and direction. In this work, average wind speed and power data recorded from 01-01-2019 to 10-12-2020 of one turbine was used, in a total of 17040 samples. This data was originally sampled in a 10-min basis and, to match the sample rate of the NWP data, it was resampled to a 1-h basis.

5 Results

The FFNN presented in Sect. 3 has been applied to the 24-h ahead wind speed forecasting of the wind turbine described in Sect. 3. The SCADA wind speed values are used as training data, and the NWP data is used as baseline.

As the NWP models can have some bias, an adjusted NWP model was considered (ANWP) by adding the average error of the NWP model to the original forecasts. This simple correction could lead to a decrease of up to 12.5% on the RMSE and to 14.9% on the mean average error (MAE) as shown in Table 1.

Table 1. Baseline models forecasting error

Baseline model	RMSE [m/s]	MAE [m/s]
NWP	1.751	1.366
ANWP	1.532	1.162

In Table 2, the results obtained on the test set with the best proposed model and baseline models are shown. One can notice that the proposed model was able to improve significantly the baseline results. The improvements were 20.6% on the RMSE, 21.7% on the MAE and 43.8% on the R2 for the NWP and 9.2% on the RMSE, 8.0% on the MAE and 12.5% on the R2 for the ANWP. Figure 1 shows the forecasts made with the proposed model, the NWP and the actual measured wind speed.

Table 2. Forecasting results of the proposed model against the baseline models

Model	RMSE [m/s]	MAE [m/s]	R2
NWP	1.751	1.366	0.584
ANWP	1.532	1.162	0.457
Proposed model	1.391	1.069	0.657

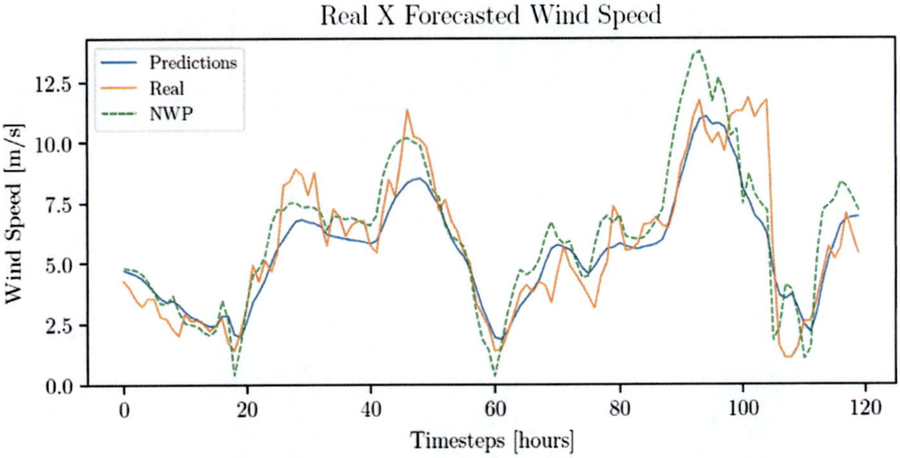

Fig. 1. Comparison of real wind speed values with the values forecasted by the proposed model and by the NWP.

Typically, the wind speed distribution can be approximated by a Weibull distribution [16]. The Weibull probability density (PDF) and cumulative functions (CDF) can be represented by the shape parameter (unitless) which usually ranges from 1 to 3 with higher values representing more variable winds; and the scale parameter which is given in *m/s* and is proportional to the mean wind speed. The PDF shows the probability of a given wind speed occur while the CDF indicates the likelihood of a speed greater than a certain value. A Weibull Distribution of the forecasts was fitted and can be seen in Fig. 2.

As it can be seen in Fig. 2, the proposed model could not generate predictions that followed the typical wind distribution, being more similar to a Gaussian Distribution and with the shape parameter above the common upper limit (3.366 vs 3). On the other hand, it was able to improve the scale parameter, which is related to the mean wind speed, by bringing it closer to the that of the true distribution for the considered turbine than the NWP outputs were (see Fig. 3*a)* and *b)*).

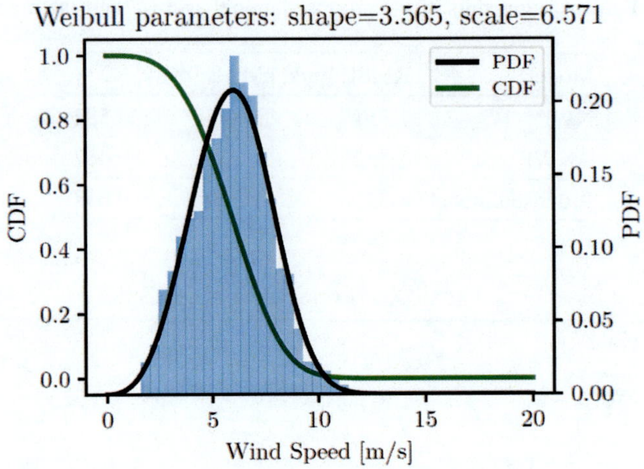

Fig. 2. Weibull Distribution of the proposed model forecasts

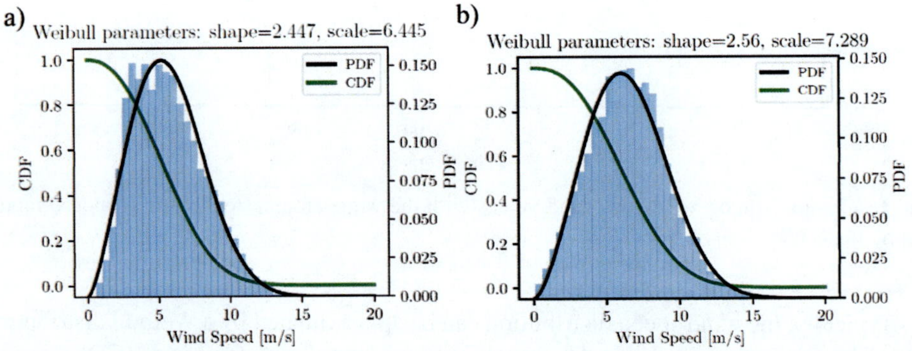

Fig. 3. *a)* Weibull Distribution of turbine wind speed; *b)* Weibull Distribution of the NWP baseline model.

6 Conclusions

The challenges and efforts to integrate renewable energy sources in the electricity grid increases the system unpredictability and variability due to the uncertain availability of these natural resources. To overcome this situation, the system must be flexible and resilient enough to cope with rapid generation and load changes and balance them at every moment.

The development and implementation of machine learning based models is cheaper and faster than other solutions that are used to address the problem of generation variability. Plenty of data from the distribution/transmission system operators, power plants, SCADA systems, market, weather forecasts, among others, is already available to be used with these models. They can provide accurate predictions of the system behaviour from energy generation to its final use and can benefit the system operators to better handle the sector issues, policy-makers to plan future actions, market agents to optimize energy tariffs and, also, producers to manage their power plants. Therefore, the employment of these models can help to achieve energy transition efficiently and are of great interest for the sector.

In this direction, this paper has proposed a novel model for wind speed forecasting based on a Feed-Forward Neural Network. The proposed model has been applied and experimented using real data from a wind farm in the south of South America, and results have shown that the proposed model achieved lower forecasting errors than the baseline approaches, including one of the most used models by wind farm operators, and an improved model with error adjustment. Thus, it can be applied by wind farm operators as it contributes to reduce wind power production uncertainty by enhancing NWP forecasts.

As future work, the analysis of the forecasting errors achieved by the proposed model will be conducted, in order to enable identifying patterns in these achieved errors with the purpose of correcting the original forecasted values and further reducing the achieved errors. Furthermore, the proposed wind speed model will be incorporated in a broader model with the goal of reaching reliable wind power generation forecasts, based on the wind speed forecasts and on the analysis of the power plan characteristic and typical generation curve.

Acknowledgements. Tiago Pinto received funding from FEDER Funds through COMPETE program and from National Funds through FCT under projects CEECIND/01811/2017 and UIDB/00760/2020. Hugo Morais was supported by national funds through FCT, Fundação para a Ciência e a Tecnologia, under project UIDB/50021/2020.

The authors would like to thank Cepel for the granted master's scholarship and for the support for the development of this research, Eletrobras for providing the data and Cepel's researchers Vanessa Guedes and Ricardo Dutra for their assistance with data analysis.

References

1. Tascikaraoglu, A., Uzunoglu, M.: A review of combined approaches for prediction of short-termwind speed and power. Renew. Sustain. Energy Rev. **34**, 243–254 (2014). ISSN: 13640321. https://doi.org/10.1016/j.rser.2014.03.033
2. Soman, S.S., Zareipour, H., Member, S., Malik, O., Fellow, L.: A review of wind power and wind speed forecasting methods with different time horizons, pp. 1–8 (2010)
3. Chang, W.-Y.: A literature review of wind forecasting methods. J. Power Energy Eng. **02**(04), 161–168 (2014). ISSN 2327-588X. https://doi.org/10.4236/jpee.2014.24023
4. Nazir, M.S., et al.: Wind generation forecasting methods and proliferation of artificial neural network: a review of five years research trend. Sustainability **12**(9) (2020). ISSN: 20711050. https://doi.org/10.3390/su12093778
5. Wang, J., Song, Y., Liu, F., Hou, R.: Analysis and application of forecasting models in wind power integration: a review of multi-step-ahead wind speed forecasting models. Renew. Sustain. Energy Rev. **60**, 960–981 (2016). ISSN: 18790690. https://doi.org/10.1016/j.rser.2016.01.114
6. European Centre for Medium-Range Weather Forecasts (ECMWF). https://www.ecmwf.int/. Accessed 28 May 2021

7. Ding, M., Zhou, H., Xie, H., Wu, M., Nakanishi, Y., Yokoyama, R.: A gated recurrent unit neural networks based wind speed error correction model for short-term wind power forecasting. Neurocomputing **365**, 54–61 (2019). ISSN: 18728286. https://doi.org/10.1016/j.neucom.2019.07.058
8. Dupré, A., Drobinski, P., Alonzo, B., Badosa, J., Briard, C., Plougonven, R.: Sub-hourly forecasting of wind speed and wind energy. Renew. Energy **145**, 2373–2379 (2020). ISSN: 18790682. https://doi.org/10.1016/j.renene.2019.07.161
9. Alonzo, B., Plougonven, R., Mougeot, M., Fischer, A., Dupré, A., Drobinski, P.: From numerical weather prediction outputs to accurate local surface wind speed: statistical modeling and forecasts. In: Drobinski, P., Mougeot, M., Picard, D., Plougonven, R., Tankov, P. (eds.) FRM 2017. SPMS, vol. 254, pp. 23–44. Springer, Cham (2018). https://doi.org/10.1007/978-3-319-99052-1_2
10. Xu, Q., et al.: A short-term wind power forecasting approach with adjustment of numerical weather prediction input by data mining. IEEE Trans. Sustain. Energy **6**(4), 1283–1291 (2015). ISSN: 19493029. https://doi.org/10.1109/TSTE.2015.2429586
11. Zheng, D., Shi, M., Wang, Y., Eseye, A.T., Zhang, J.: Day-ahead wind power forecasting using a two-stage hybrid modeling approach based on SCADA and meteorological information, and evaluating the impact of input-data dependency on forecasting accuracy. Energies **10**(12) (2017). ISSN: 19961073. https://doi.org/10.3390/en10121988
12. Bishop, C.M.: Pattern Recognition and Machine Learning. Springer, New York (2006)
13. Chen, H., Lu, F., He, B.: Topographic property of backpropagation artificial neural network: from human functional connectivity network to artificial neural network. Neurocomputing **418**, 200–210 (2020). ISSN: 0925-2312. https://doi.org/10.1016/j.neucom.2020.07.103
14. Goodfellow, I., Bengio, Y., Courville, A.: Deep Learning. MIT Press, Cambridge (2016). http://www.deeplearningbook.org
15. Geron, A.: Hands-On Machine Learning with Scikit-Learn and TensorFlow. O'Reilly Media, Sebastopol (2017). ISBN: 9781491962299
16. Burton, T., Jenkins, N., Sharpe, D., Bossanyi, E.: Wind Energy Handbook (2011). ISBN: 9780470699751

A Multi-agent Specification for the Tetris Game

Carlos Marín-Lora(✉), Miguel Chover, and Jose M. Sotoca

Institute of New Imaging Technologies, Universitat Jaume I, Castellón, Spain
{cmarin,chover,sotoca}@uji.es

Abstract. In the video game development industry, tasks related to design and specification require support to translate game features into implementations. These support systems must clearly define the elements, functionalities, and interactions of the game elements, and they must also be established independently of the target platform for its development. Based on a study for the specification of games that allows the generation of games as multi-agent systems, this work tries to check if the results can be cross-platform applied. As a case study the classic game Tetris has been used, a game whose very nature suggests that its implementation should be composed of vector and matrix data structures. The purpose is to validate the usage of a game specification based on multi-agent systems for the game's implementation on different platforms.

Keywords: Computer games · Game specification · Multi-agent systems

1 Introduction

The design and specification of video game projects is a creative process that is often performed by people with no programming knowledge. These processes aim to translate design concepts into requirements and task definition. However, there is a lack of consensus on how to establish this process [1, 2]. One of the first problems that come up are the constraints involved in designing for one platform or another depending on the required characteristics [3, 4]. In the literature, there has been a search for a framework to define the characteristics and functionalities of a game in an indirect way. For example, the field of AI research in games has contributed advances with General Game Playing (GGP) [5–8] where it describes in a way and manner that the same system is able to learn to play any game based just on the descriptions of its Game Description Language (GDL) [9, 10]. However, these methods define specification systems that require high-level technical knowledge.

An alternative approach is to consider the game elements and their behaviors as autonomous entities that solve tasks assigned to them to ensure the correct execution of a game, similar to what would be done in the real world. In other words, by presenting an analogy between the elements of a game and the autonomous agents that constitute multi-agent systems (MAS) [11]. In Marín-Lora et al. [12], a game engine able to generate games as MAS is presented. Where the game elements or agents have a set of properties and behavioral rules that allow them to interact with each other and with the

© The Author(s), under exclusive license to Springer Nature Switzerland AG 2022
K. Matsui et al. (Eds.): DCAI 2021, LNNS 327, pp. 169–178, 2022.
https://doi.org/10.1007/978-3-030-86261-9_17

social space they share, and where the definition of these behavioral rules is done by means of a formal semantics based on predicate logic. However, this work focuses on its own implementation and does not extrapolate its specification for other engines and other systems. Based on this game engine and its model, this study aims to validate the hypothesis that a game can be defined and specified as a system of interacting agents and that it is possible to implement it on different platforms. To this end, this work focuses on validating whether this model allows to define, specify and prototype a video game for multiple platforms in a fast and simple way. In addition, it is studied if it is able to define and specify the behaviors of the game elements by establishing their logics. In an intermediate way between more traditional and artistic methods such as Game Design Documents (GDD) and other more technical and advanced methods such as GDL [13]. For this purpose, the study of this work goes through the implementation of a game on three different platforms: NetLogo, GDevelop and Unity [14–16]. A MAS prototyping system, a 2D event-driven game engine and probably the most widely used game engine today, respectively. As a reference game, it is going to be used a game with a matrix nature and whose implementation a priori would not be conceived without the presence of the data structure of a matrix: the Tetris. The purpose of this case study is to validate the usage of a multi-agent specification for the implementation of games in different platforms. The article is organized as follows: Sect. 2 presents the state of the art studied for this article. Then, in Sect. 3, the data and game specification model will be presented. Subsequently, this model will be applied on the game of study in Sect. 4, and implemented on the three platforms in Sect. 5. Finally, in Sect. 6, the conclusions obtained from the realization of this work will be presented.

2 Background

As in any other software design process, in video games there are multiple paradigms or design patterns to define the code structures that manage it and to establish the logic of the behaviors of its elements [17]. Many of them are used to organize the assignment of responsibilities between elements or to define the behaviors and interactions between game objects. Some paradigms encapsulate the information needed to execute a method, perform an action or trigger an event; others are used in systems with one-to-many relationships between objects where, for example, if an object is modified, its dependent objects are automatically notified; or others that allow an object to change its behavior when its internal state is modified in a manner analogous to a state machine. Special mention should be made of iteration patterns that manage information flows. The commonest in video games is the game loop, that is, the continuous execution of the game that in each iteration processes the user interaction, the behaviors of the game elements and renders the scene in a continuous loop as long as the game state so indicates. A variant of this structure uses an auxiliary buffer as a storage method for the altered information after each iteration in order to update it in the data model at the end of the cycle and thus keep the game information intact between the beginning and the end of the iteration. And to execute logical actions, the update model is also often used, based on an update function per game element, where each element evaluates in each frame its function at local scale. It is at this point where with the evaluation of the state of the game and the autonomous execution of actions of its elements according to their internal state, the correspondence between these patterns and the MAS occurs. MASs are composed of sets of agents that interact and make decisions among themselves within a shared environment [18]. Within the shared environment, each agent has sets of properties and behavioral rules that determine its interaction with others. These agents have functions based on metrics associated with decision theory and game theory that allow them to exhibit autonomous, deliberative, reactive, organizational, social, interactive, coordinating, cooperative and negotiating behaviors [19], that have traditionally been used in autonomous robotic systems to solve real-time problems. The selection of MAS as a reference system for the specification of video games is based on the analogy between the autonomous behaviors of agents and the elements that compose games. In other words, the behaviors and interactions in these systems have correspondences with the behaviors and interactions between individuals and their environment. MAS have aspects of formal specification to define a video game in a generic way, integrating specific aspects of the game such as the game logic with its entities and their behavior rules, or the game physics with the detection and response of collisions between game elements. However, it is obvious that the relationship between video games and MAS is not new. Multiple examples relating these two categories can be found in the literature: from the construction of elements for games, the interactions between their elements or their communication and cooperation protocols [18, 20, 21]. Also for more specific purposes such as the study of role-playing game (RPG) games [22], or to define games in which a large number of people participate in areas with different influences [23]. Currently, MAS and machine learning are already incorporated in several game engines, so they are also accessible to the general public [24, 25]. For this work, the focus has been placed on the application of this paradigm on game development, and specifically on the specification of its mechanics defined by means of scripts. Scripts are routines written in a programming language that provide an abstraction layer over the systems that manage the games in the different devices, that allow modifying the state of the games without the need of recompilation, and that are usually used for the management of the behaviors and for the management of the system events [26]. Specifically, in video games, they are oriented to facilitate programming without actively thinking about optimizing the real-time execution of the game. During the last decade, the trend is towards the use of generic scripting languages, displacing languages specific to game development systems. Currently, the most widely used are C#, Python and JavaScript, and visual scripting systems such as Scratch or Unreal Blueprints [27].

3 Video Games and Specification as MAS

The goal of this work is to test if the specification of a game based on the analogy between MAS and video games is able to be implemented on different platforms. For this purpose, it is necessary to establish a formal analogy between the concepts of a game and their corresponding concepts in a MAS. In addition, the method for the definition of a game must consider the features of the game, the elements that compose it, the definition of the behaviors and the user interaction. This work uses the game specification used by Marín Lora et al. [12] for their game engine. It is based on the definition of agent proposed by M. Wooldridge [18], where an agent is a computer system located in an environment and capable of performing tasks autonomously to meet its design goals. By way of summary, some of the characteristics of the theoretical framework used for this proposal are detailed below:

- The environment to which the agents belong can be in any of the discrete states of a finite set of states $E = [\, e_0, e_1, e_2, \ldots]$.
- The environment shared by all agents has a set of properties that determine its state and can be accessed by any agent in the environment.
- The agents have generic properties (geometric, visual, physical, etc.) and they also admit new properties to perform specific tasks.
- Agents have behavioral rules for modifying the state of the environment or the state of an agent in order to meet their plans and objectives.
- Agents have a set of possible actions with the ability to transform their environment $Ac = [\, \alpha_0, \alpha_1, \alpha_2, \ldots]$.
- The run r of an agent on its environment is the interleaved sequence of actions and environment states r: $e_0 \rightarrow \alpha_0\ e_1 \rightarrow \alpha_1\ e_2 \rightarrow \alpha_2 \ldots e_{u\text{-}1} \rightarrow \alpha_{u\text{-}1}\ e_u$.
- The set of all possible executions is R, where R^{AC} represents the subset of R that ends with an action, and R^E represents the subset of R that ends with a state of the environment. The R members are represented as $R = [r_0, r_1, \ldots]$.
- The state transformation function introduces the effect of an agent's actions on an environment τ: $R^{AC} \rightarrow \wp(E)$ [28].

In order to transfer these concepts to a video game and to any support, it is necessary to define the general characteristics of the game and those of its elements such as those of an environment and its agents, respectively. Considering that there must be analogous functions and attributes for each element, regardless of the limitations or features of the game engine or software environment selected for its implementation.

Following the reference model, the rule specification is structured using first-order logical semantics [29] based on two predicates: an IF condition and a DO action. Where each predicate executes calls to actions α of the system or evaluates arithmetic and Boolean expressions. The predicates specify the logic of the game so that the tasks to be performed by an agent are organized in predicate formulas where their elements can have the following predicate structures:

- **Action structure**: Composed of an atomic element including a single predicate literal DO.
- **Conditional structure**: Generated from the structure of the IF-THEN-ELSE rules [30].

$$(\text{IF} \rightarrow \varphi) \wedge (\neg\text{IF} \rightarrow \theta)$$

where IF is a conditional literal predicate, and where φ and θ are sequences of new predicates that will be evaluated if the condition is met or if it is not met, respectively. The conditional predicate represents the evaluation element of a condition in the decision making process. Where the evaluation of the condition is based on the result of a logical expression that values the relationship between system entities. This logical expression may contain arithmetic expressions composed from system properties, game or agent properties, mathematical functions and numerical values.

$$\text{IF}\,(\textit{expression})$$

Based on the evaluation of these expressions, the logical elements determine the need for a game agent to perform an action α in the game. An α-action is defined as a behavior to be performed by an agent, and are formalized as non-logical function elements that can handle parameters such as arithmetic expressions. The set of actions is based on the create, read, update, and delete (CRUD) operations of information persistence [31] applied to the game properties and its agents.

- **Create**: Creates a new agent, as a copy of an existing agent.
- **Read**: Reads the information of a game or a property of a game object. The syntax *agent.property* is used to read this information.
- **Update**: Modifies the value of a property of a set or an agent. The new value is determined from the evaluation of an arithmetic expression.
- **Delete**: Removes an agent from the game.

An example of these rules and the game specification could be presented as follows:

AG1:

- **Properties**: { A: *true*, B: *1.00*, C: "*Hi!*"}
- **Scripts**: { IF(A) $\rightarrow$ DO(B $=$ B $+$ 1) $\wedge$ DO(C $=$ "*My name is AG1*")}

AG2:

- **Properties**: { A: *false*, B: *-1.00*, C: "*What is your name?*"}
- **Scripts**: { IF(B $\leq$ 0) $\rightarrow$ DO(C $=$ AG1.C)$\wedge\neg$IF(A) $\rightarrow$ DO(*delete* AG1)}

From this model, the specification system designed must be able to define the behaviors of the elements that make up the sets in a general way.

4 Case Study: Tetris

The game to be used as a case study is Tetris. This classic logic arcade game was originally designed and implemented by Aleksei Pazhitnov in the Soviet Union and published in 1986 [32]. The game consists of falling pieces composed of four blocks in different configurations. The goal of the game is to stack the pieces at screen bottom so that the accumulation of pieces does not reach the screen top. To avoid this, the player must position and rotate the pieces as they fall so as to complete as many horizontal rows as possible. Each time a row is completed, it vanishes and the blocks above it fall. Each time a piece is stacked, a random new piece appears from the screen top. There are seven variations of pieces in the game, each with a specific shape and named O, L, J, T, I, S, Z and T. Figure 1 shows a representation of them in the same order in which they have been described from left to right.

Fig. 1. Diagram of game piece shapes

The data model of this game starts from the pieces, composed of four blocks arranged in a predetermined configuration. In addition, the player can perform geometric transformations on them to change their position and orientation. Finally, the game must eliminate blocks when they complete a row. Therefore, the types of agents needed for this implementation are three: the piece, the block and the checker. A representation of these three agents can be seen in Fig. 2.

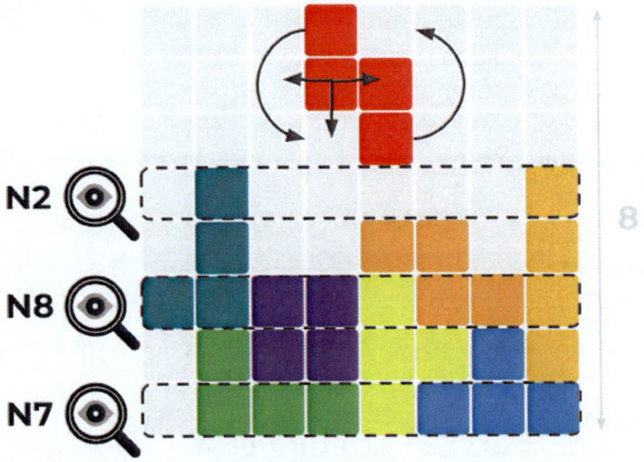

Fig. 2. Diagram of game mechanics

Piece Agent: Composed of four block agents in a prearranged layout. There is only one piece in the game at a time: the falling piece. While falling, the player can modify its position in a unit left, right and down using the corresponding arrow keys. In addition, he/she can rotate its orientation to the left and right with the L and R keys, respectively. As soon as it comes to rest with blocks already placed or with the background, it is removed but the blocks that compose it are kept. These blocks will remain in their position until their line is completed or until the end of the game.

- **Properties**: { resting: *false*}
- **Movement script**: { ¬IF(resting) → (DO(y = y + *1*) ∧ IF(Game.KeyArrowLeft) → DO (x = x - *1*) ∧ IF(Game.KeyArrowRight) → DO (x = x + *1*) ∧ IF(Game.KeyArrowDown) → DO (y = y + *1*) ∧ IF(Game.KeyR) → DO (angle = angle - *90*) ∧ IF(Game.KeyL) → DO(angle = angle + *90*)}
- **Rest script**: { IF(collisionRest) → (DO(resting = *true*) ∧ DO(delete))}

Block Agent: hey initially compose a piece and move with it. By default, they are considered to have a dimension of one unit. When the piece goes to rest state, they are unlinked from it and remain static in their waiting position. It has a property to store the information of the row for the moment in which it is at rest. If they come across a check agent in elimination mode, they must communicate to the piece above them that it has to move down one position, and then it must be eliminated from the game.

- **Properties**: { resting: *false*, row = *-1*}
- **Set script**: { IF(resting) → DO(row = y)}
- **Destroy script**: { IF(collisionCheck) → IF(check.deleteMode) → ((IF(collisionBlockUp) → IF(blockUp.y > y) → DO(blockUp.y = blockUp.y + *1*) ∧ DO(delete))}

Check Agent: In each of the rows there is a controller agent that checks the number of blocks occupying its row. When activated, it runs from left to right through its row checking if each position is occupied by a block. If when it reaches the end of the row its number is equal to the number of existing columns, it must return in the opposite direction informing all the blocks in its row that they must be destroyed, and it must send a message to all the checker agents in the rows above it to move the blocks at rest one position down.

- **Properties**: { count: *0*, deleteMode: *false*}
- **Forward script**: { IF(x ≤ Game.width) → DO(x = x + *1*) ∧ ¬IF(x ≤ Game.width) → DO(x = *0*)}
- **Backward script**: { IF(x > *0*) → IF(deleteMode) → DO(x = x - *1*)}
- **Check script**: { IF(collisionBlock) → DO(count = count + *1*)}
- **Clear script**: { IF(count = = Game.width) → DO(deleteMode = *true*)}

5 Results and Discussion

From this specification, a version of the game has been successfully implemented on the three intended systems: NetLogo, GDevelop and Unity. All three have respected the game description and the game definition as MAS. The most notable particularities of these implementations reside mainly in the graphical interface available and in some features in the way of composing the game logic. In the case of NetLogo, and in contrast to the other two systems, there is no graphical editor to create the initial spatial layout of the game interactively and it has been necessary to create this initial configuration through its API and its scripting system based on the turtle language. Moreover, since this system is not intended to generate games, its graphic quality has been reduced to regions of colored pixels. This system has the particularity of having "broadcast" functions for one type of agent, so that communication could occur in a more direct way. However, the model has been followed and broadcast through the use of auxiliary variables. At the logical level, the major particularity has been the rotation of the pieces. It has been chosen to follow the methodology used by an example of the game in its resource library, where the rotation occurs through the exchange of positions between pieces from the definition of a central block agent defined as block 0. In the case of GDevelop, the game has been arranged in its web editor and the logic has been implemented through its logic system based on event management. In contrast to the previous case, the graphical level of the system has allowed to generate more visual forms. The most outstanding particularity of this system in terms of logic has been the communication between agents. For example, after collision events it has been necessary to create auxiliary variables in the general properties of the game and to subscribe potentially interested agents to these variables. Finally, Unity is the most powerful of the three environments. It has made it possible to compose the game and its specification from its editor and its scripting system in the C# language. The particularities of this implementation are very similar to those found in GDevelop, where communications have been made through game variables and each of the agents checked its status after each iteration of its Update function.

6 Conclusions

The work presented in this paper aims to validate the hypothesis that a game can be defined and specified as a system of interacting agents and that it is possible to implement it on different platforms. For this purpose, a model for the specification of games based on a game engine created to generate games as multi-agent systems has been taken as a starting point. From which it has been studied and tested whether the specification of a game with this model can be implemented on different platforms. For its validation, the classic game Tetris, a game that by nature should be based on vector and matrix structures, has been used as a reference. Finally, the game has been defined and specified according to the reference model obtaining a total of three different agent types for the game composition. With this specification, the same system has been implemented in three different platforms with satisfactory results. With this, it can be said that the starting hypothesis has been successfully validated and the objectives of this work have been met. As a future work, the extension of the specification system is being considered through the definition of a formal language that would allow the specification and programming of games following this same model based on MAS and based on first-order logic.

Acknowledgements. This work has been funded by the Ministry of Science, Innovation and Universities (PID2019-106426RB-C32/AEI/https://doi.org/10.13039/501100011033, RTI2018-098651-B-C54) and by research projects of the Universitat Jaume I (UJI-B2018-56, UJI-B2018-44, UJI-FISABIO2020-04).

References

1. Anderson, E.F., Engel, S., Comninos, P., McLoughlin, L.: The case for research in game engine architecture. In: Proceedings of the 2008 Conference on Future Play: Research, Play, Share, pp. 228–231 (2008)
2. Ampatzoglou, A., Stamelos, I.: Software engineering research for computer games: a systematic review. Inform. Softw. Technol. **52**(9), 888–901 (2010)

3. Anderson, E.F., et al.: Choosing the infrastructure for entertainment and serious computer games—a whiteroom benchmark for game engine selection. In: 2013 5th international conference on games and virtual worlds for serious applications (VS-GAMES), pp. 1–8. IEEE (2013)
4. BinSubaih, A., Maddock, S., Romano, D.: A survey of game portability. University of Sheffield, Tech. Rep. CS-07-05 (2007)
5. Genesereth, M., Love, N., Pell, B.: General game playing: overview of the AAAI competition. AI Mag. **26**(2), 62 (2005)
6. Perez-Liebana, D., Samothrakis, S., Togelius, J., Schaul, T., Lucas, S.M.: General video game ai: competition, challenges and opportunities. In: Thirtieth AAAI Conference on Artificial Intelligence (2016)
7. Thielscher, M.: A general game description language for incomplete information games. In: Twenty-Fourth AAAI Conference on Artificial Intelligence, July 2010
8. Thielscher, M.: The general game playing description language is universal. In: Twenty-Second International Joint Conference on Artificial Intelligence (2011)
9. Love, N., Hinrichs, T., Haley, D., Schkufza, E., Genesereth, M.: General Game Playing: Game Description Language Specification (2008)
10. Ebner, M., Levine, J., Lucas, S.M., Schaul, T., Thompson, T., Togelius, J.: Towards a Video Game Description Language (2013)
11. Dorri, A., Kanhere, S.S., Jurdak, R.: Multi-agent systems: a survey. IEEE Access **6**, 28573–28593 (2018)
12. Marín-Lora, C., Chover, M., Sotoca, J.M., García, L.A.: A game engine to make games as multi-agent systems. Adv. Eng. Softw. **140**, 102732 (2020)
13. Schiffel, S., Thielscher, M.: A multiagent semantics for the game description language. In: Filipe, J., Fred, A., Sharp, B. (eds.) ICAART 2009. CCIS, vol. 67, pp. 44–55. Springer, Heidelberg (2010). https://doi.org/10.1007/978-3-642-11819-7_4
14. Wilensky, U.: NetLogo. http://ccl.northwestern.edu/netlogo/. Center for Connected Learning and Computer-Based Modeling, Northwestern University, Evanston, IL (1999)
15. Gdevelop: https://gdevelop-app.com/. Last accessed 28 May 2021
16. Unity: https://unity.com/. Last accessed 28 May 2021
17. Nystrom, R.: Game Programming Patterns. Genever Benning (2014)
18. Wooldridge, M.: An Introduction to Multiagent Systems. John Wiley & Sons (2009)
19. Silva, C.T., Castro, J., Tedesco, P.A.: Requirements for Multi-Agent Systems. WER **2003**, 198–212 (2003)
20. Poslad, S.: Specifying protocols for multi-agent systems interaction. ACM Trans. Autonom. Adaptive Syst. **2**(4), 15 (2007). https://doi.org/10.1145/1293731.1293735
21. Marin-Lora, C., Chover, M., Sotoca, J.M.: Prototyping a game engine architecture as a multi-agent system. In: 27th International Conference in Central Europe on Computer Graphics, Visualization and Computer Vision (2019)
22. Barreteau, O., Bousquet, F., Attonaty, J.M.: Role-playing games for opening the black box of multi-agent systems: method and lessons of its application to Senegal River Valley irrigated systems. J. Artif. Soc. Soc. Simul. **4**(2), 5 (2001)
23. Aranda, G., Trescak, T., Esteva, M., Rodriguez, I., Carrascosa, C.: Massively multiplayer online games developed with agents. In: Pan, Z., Cheok, A.D., Müller, W., Chang, M., Zhang, M. (eds.) Transactions on edutainment vii. LNCS, vol. 7145, pp. 129–138. Springer, Heidelberg (2012). https://doi.org/10.1007/978-3-642-29050-3_12
24. Juliani, A., et al.: Unity: A general platform for intelligent agents. arXiv Preprint, arXiv:1809.02627 (2018)
25. Chover, M., Marín, C., Rebollo, C., Remolar, I.: A game engine designed to simplify 2D video game development. Multimedia Tools and Applications **79**(17–18), 12307–12328 (2019). https://doi.org/10.1007/s11042-019-08433-z

26. Anderson, E.F.: A classification of scripting systems for entertainment and serious computer games. In: 2011 Third International Conference on Games and Virtual Worlds for Serious Applications, pp. 47–54. IEEE (2011)
27. Rebollo, C., Marín-Lora, C., Remolar, I., Chover, M.: Gamesonomy vs scratch: two different ways to introduce programming. In: 15th International Conference on Cognition And Exploratory Learning In The Digital Age (CELDA 2018). Ed. IADIS Pres (2018)
28. Fagin, R., Moses, Y., Halpern, J.Y., Vardi, M.Y.: Reasoning about knowledge. MIT Press (2003)
29. Brachman, R.J., Levesque, H.J., Reiter, R. (eds.): Knowledge Representation. MIT Press (1992)
30. Karplus, K.: Using if-then-else DAGs for multi-level logic minimization. Computer Research Laboratory, University of California, Santa Cruz (1988)
31. Daissaoui, A.: Applying the MDA approach for the automatic generation of an MVC2 web application. In: 2010 Fourth International Conference on Research Challenges in Information Science (RCIS), pp. 681–688. IEEE (2010)
32. Wilensky, U.: NetLogo Tetris model. Center for Connected Learning and Computer-Based Modeling, Northwestern University, Evanston, IL. http://ccl.northwestern.edu/netlogo/models/Tetris (2001)

Service-Oriented Architecture for Data-Driven Fault Detection

Marta Fernandes[1(✉)], Alda Canito[1], Daniel Mota[1], Juan Manuel Corchado[2], and Goreti Marreiros[1]

[1] GECAD - Research Group on Intelligent Engineering and Computing for Advanced Innovation and Development, Polytechnic of Porto (ISEP/IPP), Porto, Portugal
{mmdaf,alrfc,drddm,mgt}@isep.ipp.pt

[2] BISITE Research Centre, University of Salamanca (USAL), Salamanca, Spain
corchado@usal.es

Abstract. Predictive maintenance approaches result in shorter interventions, less downtime and less maintenance and production costs, leading to more sustainable manufacturing and increasing earnings. Machine learning and data mining approaches are now popular in this field, producing data-driven models that can provide insights regarding equipment behavior, identify anomalies, and generate predictions that can be used to support decision-making processes. In this paper, we propose a service-oriented architecture that adapts a predictive maintenance reference architecture to the domain of the flexible packaging market. Focusing on the data analytics services required for predictive maintenance, the proposed architecture was assessed by conducting a case-study using real-world data. To detect faults in the company's coextrusion film blowing machines, we devised a predictive maintenance methodology whereby the isolation forest algorithm is used to detect potentially anomalous data points and the anomaly threshold is defined by computing the interquartile range of anomaly scores. To compensate for the fact that single anomalies are unlikely to be of significance, when deploying the isolation forest model in real time the percentage of anomalies detected within a given time interval is monitored. An alarm is issued by the system should that percentage exceed a specific cutoff. Experimental results show the described approach is successful in distinguishing anomalous data from normal data in non-stationary time-series.

Keywords: Service-oriented architecture · Fault detection · Predictive maintenance · Machine learning

1 Introduction

Companies are becoming more and more aware of the importance of the continuous monitorization of equipment condition for the application of preventive and predictive maintenance (PdM) approaches. Equipment failure not only results in downtime, and therefore in higher production costs, but may also potentially damage articles in production [1]. By anticipating potential issues in the equipment, PdM approaches result in

© The Author(s), under exclusive license to Springer Nature Switzerland AG 2022
K. Matsui et al. (Eds.): DCAI 2021, LNNS 327, pp. 179–189, 2022.
https://doi.org/10.1007/978-3-030-86261-9_18

shorter interventions and less downtime, especially when compared with interventions performed only after faults have occurred [2, 3]. Replacing deteriorating components before major faults occur also has the upside of reducing the probability of future failure and increasing the intervals between necessary interventions.

Predictive maintenance relies on the detection and prediction of failures in equipment through the analysis of its past behavior: it is a form of condition-based maintenance in which the evolution of a set of parameters can determine the current, real condition of the equipment, and be used to predict future states [4–6]. Machine learning and data mining techniques are now popular approaches in this field, producing data-driven models that can provide insights regarding equipment behavior, identify anomalies, and generate predictions that support the decision-making processes within companies [4, 7]. The application of these techniques is particularly relevant in complex scenarios, with large volumes of data and parameters [1].

The work proposed in this paper has been developed in the scope of project PIANiSM, which aims to build an end-to-end predictive maintenance platform suitable for different industrial domains. To achieve that, the platform's reference architecture has been designed to be modular and incorporate not only fundamental components, but also services that might be useful in specific domains. The high-level reference architecture is composed of four layers: 1) data acquisition layer, 2) data pre-processing layer, 3) model development layer and 4) applications' layer.

This paper describes the adaptation of layers two and three of the reference architecture to the domain of flexible packaging, specifically the production of flexible films. To address this use-case, we propose a service-oriented architecture (SOA) and a predictive maintenance strategy to detect faults in plastic coextrusion machines. As an architectural style that promotes loose coupling between components, SOA provides an adequate model to define the implementation of PIANiSM's reference architecture. Additionally, the proposed PdM methodology was developed to handle the constraints created by non-stationary time series data, the absence of labelled data and the need to detect faults in real-time.

The rest of this document is organized as follows: Sect. 2 presents some related concepts. Section 3 describes the system architecture and Sect. 4 presents a case-study of the developed system. Finally, Sect. 5 provides the concluding remarks and directions for future work.

2 Background

2.1 Service-Oriented Architecture

The service-oriented architecture can be seen as a conceptual model where business-aligned software components provide services to other components, communicating over a network [8]. It is an architectural style that defines how loosely-coupled software components should be developed and interact [8, 9]. Software components in a service-oriented architecture represent repeatable business activities and are developed according to the principle of separation of concerns, i.e., the implementation details of one component are hidden from other components. As a result, services can be used and updated independently, as well as combined and reutilized, making for resilient and highly flexible systems [8, 9].

In a SOA, service providers publish a description of the service they supply, and service consumers invoke those services. Services may be exposed and invoked through a service registry (optional), which is typically an Enterprise Services Bus (ESB) within the enterprise, but messages may also be exchanged across the Web when dealing with external services. The service consumer can also access the service description directly, however, depending on a system's complexity, implementing point to point connections might be laborious and challenging to maintain [8]. Although architectures are independent of specific technologies, in SOA the interaction between services is usually done using standard network protocols such as SOAP or RESTful HTTP [9].

A concept related to SOA is that of microservices. Consensus has yet to be reached on whether microservices represent an independent architectural style or are an implementation approach to SOA, but many experts seem to support the latter standpoint [10]. Whereas SOA defines the integration of business-aligned components, microservices apply SOA's principles and patterns at the application level [10].

2.2 Isolation Forest

The isolation forest algorithm was proposed in 2008 by Fei Tony Liu, Kai Ming Ting and Zhi-Hua Zhou [11]. As the name suggest, an isolation forest is a collection of tree structures that isolate anomalous data points. Unlike most anomaly detection algorithms, the isolation forest algorithm does not create a profile of normal data and define as anomalous the data points that do not match that profile. Instead, it relies on the fact that anomalous points are scarce and distinct to separate them from normal data by partitioning the feature space. Anomalous data points are easier to isolate than normal data points that are usually more numerous and lie closer together in the feature space [11].

Each isolation tree (iTree) in an isolation forest is built by randomly selecting a feature and randomly choosing a split value between the minimum and maximum values of the selected feature. Following this procedure, a data sample is partitioned recursively until each external node contains a single instance, or all the instances at one node have the same values. The number of splits necessary to isolate a data point is equivalent to the path length from the iTree's root node to the external node containing that data point. Since an isolation forest is an ensemble of iTrees, the expected path length is the average length across all iTrees. Since the isolation forest was developed for the purpose of detecting anomalies, the algorithm assigns to each observation an anomaly score that is defined as:

$$s(x, n) = 2^{-\frac{E(h(x))}{c(n)}} \tag{1}$$

where $h(x)$ is the path length of observation x, $E(h(x))$ is the average of path lengths of x from each iTree, and $c(n)$ is the average path length of unsuccessful search in a Binary Search Tree [11]. A score s very close to 1 means an observation is an anomaly, a score much smaller than 0.5 indicates the observation is likely normal, and if the scores of all observations are approximately 0.5 then no anomalies exist in the data [12]. Additionally, to handle clustered anomalies and provide results with different levels of granularity a limit *hlim* can be set to the path length, effectively defining an anomaly threshold [12].

The isolation forest algorithm only requires the definition of two parameters: the number of iTrees and the subsampling size. Since iTrees do not need to isolate all the normal data points, they can be built by discarding most of the training data. In fact, the isolation forest produces better results when the sample size is kept small because a larger sample size makes it harder to isolate anomalies due to problems of masking (existence of too many anomalies) and swamping (normal instances are identified as anomalies) [11]. Because of this, the isolation forest algorithm has linear time complexity and low memory requirements, making it well suited to handle large volumes of data [11, 12].

3 System Architecture

The architecture conceived for the flexible packaging market use case employs a service-oriented approach where there are service consumers that require the services of the providers to achieve their functional goal in the pipeline defined for the predictive maintenance platform. Three main service providers are identified according to the reference architecture for the PIANiSM platform: 1) the data modelling service; 2) the pre-processing service and 3) the data access service. These services are independently deployed, maintained, and communicate with each other using standard technologies that operate over an IP network.

Service consumers range from dashboards and UI controls that expose features of the platform to the end-users of the system, to back-office systems that provide configuration and fine-tuning to the overall system, but a service provider may itself consume other services, e.g., the pre-processing service is a service provider that also consumes the services exposed by the data access service.

The integration of these services has been achieved using the Zato Framework [13]. The Zato Framework is a highly scalable Python-based enterprise integration platform. Zato allows services to be connected over common technologies such as REST, SOAP, WebSockets, AMQP, etc. Using Zato as a middleware, the different services can be called by a single name, defined during the configuration of the services. As such, a service's business logic is completely encapsulated, and any change done to it is performed seamlessly without impacting the service consumer.

As previously mentioned, the architecture features three main service providers. Figure 1 presents the service providers deployed within the Zato middleware, how they relate to each other, the API services that each of them has access to, and the consumers that may use their services.

The data access service provides data access functionalities to the system. The consumers are given access to both historical and real-time data from the machines that are being monitored. Each machine has its own set of variables or features that correspond to different signals, sourced from the various sensors that are equipped in each machine. These sensors are built-in by the manufacturer in the machine but may also be external sensors installed to enrich the acquisition of relevant information that describes the behavior of the machine. The historical data is provided by a RESTful API that, given the identification of the machine and the different sensors as well as a time window, returns

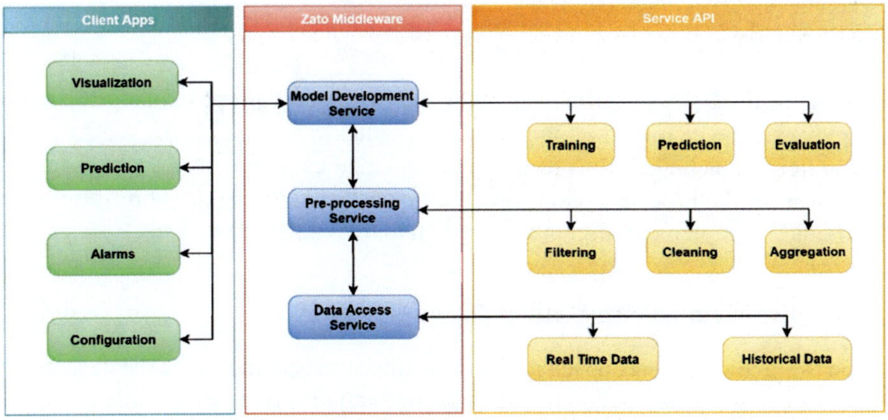

Fig. 1. Architecture of the PIANiSM platform.

the sensor data collected during the required time frame. Realtime data is provided by a data streaming service implemented within the data access service, for each available variable of each machine.

The pre-processing service focuses on the data manipulation requirements of a predictive maintenance application. At this stage, several component services are available for handling the data acquired from the data access service, such as synchronization, aggregation, interpolation, etc. Among other purposes, the processes of this service may provide performance optimization for the subsequent modelling pipeline.

The modelling service is used to build and evaluate machine learning models that predict and detect faults in the machines. This service uses two different endpoints from the pre-processing service, each of them using a different approach to the learning pipeline. One is the endpoint to the offline learning component that processes the historical data used to train offline models. The second is the endpoint to the online learning component, which handles smaller volumes of real-time data to build online models.

The offline learning component builds models according to a specific predictive maintenance strategy, namely regression models for prediction of an equipment's remaining useful life (RUL), anomaly detection algorithms for fault detection, or classification models to predict the occurrence of failures within a given time frame. This component includes behaviors dedicated to the tasks of training, evaluating, or fine-tuning the models.

The online learning component serves the purpose of allowing the system to learn continuously from the data as it arrives in real-time. Machine learning models adequate for data stream learning are implemented in this component and retrained either sample by sample or after *n* samples have been collected and potentially aggregated. As new data arrives, there is the possibility that the model might become outdated and not fit the nature of the changes that happened overtime. As such, a strategy is defined to determine if there is the need to retrain and update the model with more recent data.

4 Case Study

A prototype of the PdM system developed according to the proposed architecture has been deployed in the factory of a company that operates in the flexible packaging market, producing flexible technical films for the food and medical industries. The present case-study focuses on the procedure developed to detect faults in coextrusion film blowing machines and its implementation in the predictive maintenance system.

4.1 Predictive Maintenance Methodology

Although the architecture proposed in Sect. 3 is meant to accommodate the implementation of different predictive maintenance strategies and different modes of learning (i.e., offline, and online learning), the proof-of-concept described in this paper focuses on the detection of faults using unsupervised learning methods. Specifically, the detection of faults in the industrial machines is approached as an anomaly detection task and performed using the isolation forest algorithm.

During the training phase, an isolation forest is trained for each univariate time series obtained from the sensors that monitor the machines. Considering time series $\{X_t\}, t \in T$, the isolation forest is fit to a sliding window A of size $W1$ (the size can be defined as a time period or as number of instances). After fitting the isolation forest to the data, the anomaly score for every instance is calculated. However, even though the score can be used to identify instances as anomalies, an anomaly threshold must be carefully defined since it is affected by the business's needs (e.g., tradeoff between false positives and false negatives) and the characteristics of the data. Although an anomaly threshold can be defined by specifying the *hlim* parameter of the isolation forest algorithm, when deploying the model in real time the distribution of the data might change, which will change the anomaly scores. If the anomaly threshold remains the same as the anomaly scores change, it might result in more false positives or false negatives. To address this issue, instead of setting a threshold based on the *hlim* parameter, the anomaly threshold is based on the interquartile range (IQR) of the anomaly scores, with the lower and upper limits defined as such:

$$Lower\ limit = Q1 - k \times IQR \tag{2}$$

$$Upper\ limit = Q3 + k \times IQR \tag{3}$$

In Eqs. 2 and 3, $Q1$ is the first quartile of the data contained in window A, $Q3$ is the third quartile and the IQR is given by:

$$IQR = Q3 - Q1 \tag{4}$$

Parameter k is usually 1.5, which represents 2.7 standard deviations from the mean, that is, the data points with anomaly scores larger or smaller than 2.7 standard deviations from the mean are considered anomalies. However, k controls the sensitivity of the decision interval and can therefore be adjusted to balance the number of false negatives and false positives.

A single anomaly is unlikely to have any significance, it is the accumulation of anomalies over a given time period that indicates the possibility of an impending malfunction. As such, when deploying the isolation forest model in real time, the system monitors the percentage of anomalies detected within a sliding window B of size $W2$ of the most recent data. If the percentage of anomalies within the window exceeds a cutoff X the system issues a warning that a failure may be imminent. To continuously monitor the incoming data, window B is moved by a given length $S2$.

To handle the time-varying properties of production processes, the model must be regularly updated with the newest data. To achieve that and taking advantage of the isolation forest's linear time and low memory requirements, window A is also moved by a given length $S1$ and the model is updated. Parameters $W1$, $W2$, X, $S1$ and $S2$ should be defined according to the business's requirements.

4.2 Experimental Results

The data used in this case-study comes from a 9-layer coextrusion film blowing machine. Historical data relative to each of the nine extruders has been collected from the machine's built-in sensors at a rate of twelve times per minute (every five seconds). Table 1 presents a list of the data features available for each extruder. The different extruders produce different layers of plastic material, each with a particular thickness. That thickness is recorded by the 'layer thickness' feature and measured as a percentage of the make-up of the final product. The 'instantaneous weight' records the weight of bulk material in an extruder's hopper. The 'melt pressure' and 'melt temperature' refer to the pressure and temperature inside the extrusion barrel, respectively, whereas the 'motor rotation' feature registers the speed of the drive motor in revolutions per minute (RPM). Finally, the 'throughput' refers to the rate at which material exits the extruder head.

Table 1. Monitored parameters.

Data feature	Unit of measure
Layer thickness	Percentage
Instantaneous weight	kg
Melt pressure	bar
Motor rotation	RPM
Melt temperature	°C
Throughput	kg/h

Prior to building an isolation forest, the data had to be pre-processed to correct some issues and transform it to a more usable format. Namely, it was first necessary to clean the data, performing tasks such as removing corrupt data introduced by erroneous sensor readings and removing duplicate entries. The time series were also synchronized, since there were some discrepancies in the timestamps of the different variables, and linear interpolation was used to fill in missing values. Additionally, the data was undersampled to reduce its frequency from 5-s intervals to 5-min intervals. This change in frequency reduced the volume of data without significantly changing the expressiveness of the data.

After the preparation phase, an isolation forest was fit to each time series following the procedure described in Sect. 4.1. Each model was trained using a sample of one month of data and the most adequate number of iTrees and the subsampling size were determined by building several models with different parameter values. After fitting the models, the anomaly scores were calculated, as were the respective anomaly thresholds using the IQR. Since labelled data was not available, the models' performance had to be assessed visually, i.e., the detection results of each model were plotted and compared to decide which combination of parameter values yielded the best results for each variable.

Two other anomaly detection techniques, specifically the density-based spatial clustering of applications with noise (DBSCAN) algorithm and a low-pass filter combined with a modified version of Z-score, were applied to the data to compare their results with the ones obtained using the isolation forest algorithm with the IQR. DBSCAN is a density-based clustering algorithm proposed by Ester *et al.* [14] that is commonly used for anomaly detection tasks. It was chosen for application in this case study because it does not require the number of clusters to be defined as a parameter. The low-pass filter approach consisted in computing a moving average of the time series data and flagging as anomalies the data points whose modified Z-scores were greater than 3.5 (more than 3.5 standard deviations above or below the moving average). A modified version of Z-score that uses the median and the median absolute deviation (MAD) instead of the mean and the standard deviation was used in this approach since, unlike the mean, the median is robust to outliers. An instance is usually flagged as an anomaly if its Z-score is greater than three, but when applying the modified Z-score 3.5 is the recommended value [15].

Figure 2 shows the anomalies detected in one month of melt temperature data after applying the three anomaly detection techniques. Detecting anomalies in this data is particularly challenging because, since different products are manufactured in the same production line, the data is non-stationary. Changes in machine configurations and production materials are reflected in the sensor data and can be mistaken for anomalies when in fact they represent normal manufacturing processes. Despite that, as can be observed in Fig. 2, all three techniques perform reasonably well. The low-pass filter with modified Z-score detects anomalous points quite well, but it also marks normal data close to those points as anomalies. The results of DBSCAN and the isolation forest plus IQR are similar, but DBSCAN seems to detect them with greater precision. The isolation forest approach outputs more false positives than DBSCAN, particularly in the case of shorter manufacturing processes. However, because of its linear time and low memory requirements the isolation forest is well suited for being deployed in real-time, particularly when combined with the IQR, as explained in Sect. 4.1. On the contrary, DBSCAN is a less viable option for real-time deployment because, given the non-stationarity of the data, the model needs to be updated regularly and automatically. The quality of DBSCAN's output depends on finding appropriate values for its input parameters but determining them automatically poses a significant challenge.

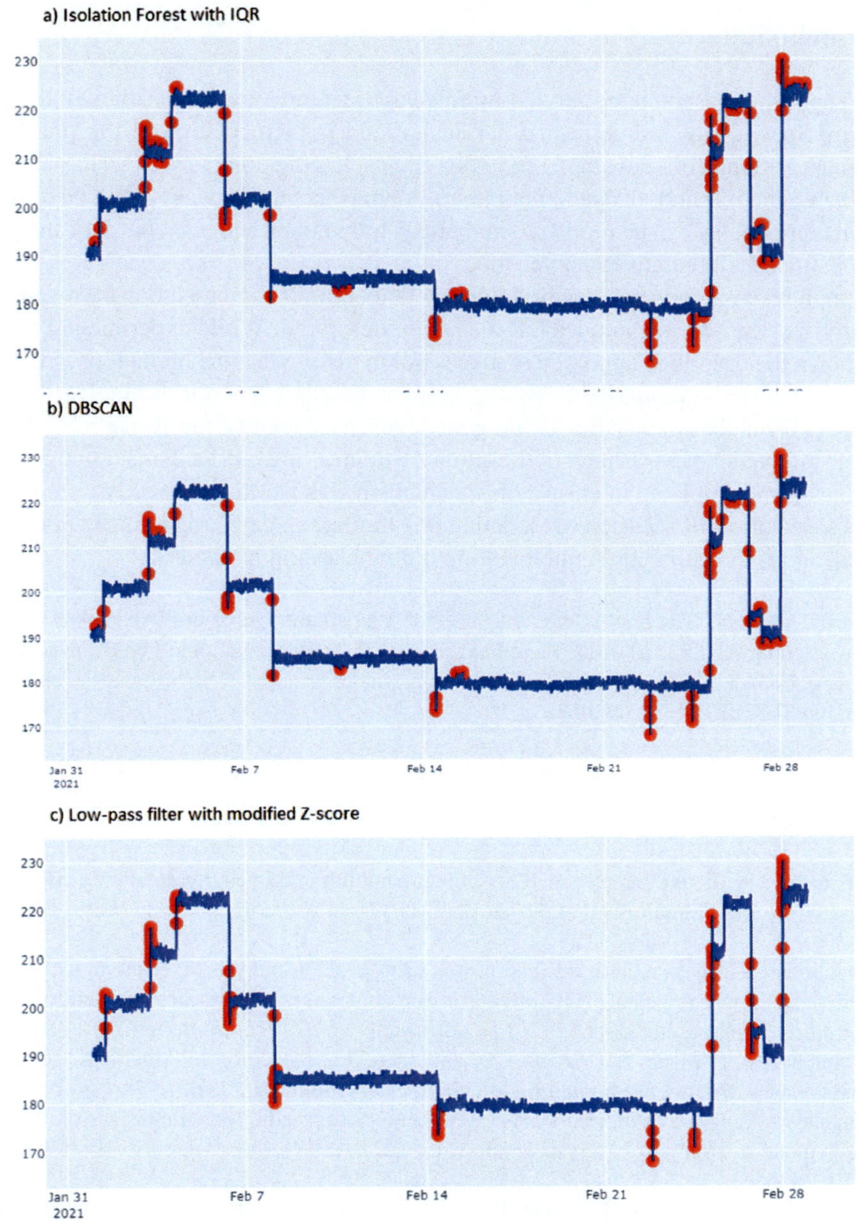

Fig. 2. Anomalies detected in one month of melt temperature data from extruder A by a) isolation forest with $n\_estimators = 300$ and $subsampling = 2000$ + IQR with $k = 2$, b) DBSCAN with $minPts = 8$ and $\varepsilon = 0.2$ and c) low-pass filter + modified Z-score.

The isolation forest models plus IQR were then deployed in real time and used to detect faults in 24 h of the latest data. Whenever the number of anomalies in the 24-h time window exceeds 25%, the system issues an alarm. The window is moved every hour to incorporate the latest data and discarding the oldest one. Similarly, the isolation forest is updated every 15 days with a new batch of one month of data.

5 Conclusion

In this paper, we presented the service-oriented architecture of a predictive maintenance platform. Specifically, we described the design and implementation of the predictive maintenance framework developed for a flexible packaging company. SOA has allowed us to create a flexible and easily maintainable system where the services provided by independent components can be reutilized and orchestrated to provide diverse data analytics functionalities for predictive maintenance.

A predictive maintenance methodology to detect faults in non-stationary data using unsupervised learning techniques has also been described. While experimental results have shown the described approach is successful in distinguishing anomalous data from normal data in non-stationary time-series, since no labelled data was available it was not possible to formally evaluate the learning models. Consequently, the absence of labelled data also affected the assessment of the anomaly monitoring mechanism, although some anomalies were simulated to ensure the mechanism was working correctly.

Future efforts will focus on overcoming this limitation, detecting faults in multivariate data, and researching and implementing online learning methods.

Acknowledgements. The present work has been developed under project PIANiSM (EUREKA – ITEA3: 17008; ANI|P2020 40125) and has received Portuguese National Funds through FCT (Portuguese Foundation for Science and Technology) under project UIDB/00760/2020 and Ph.D. Scholarship SFRH/BD/136253/2018.

References

1. Selcuk, S.: Predictive maintenance, its implementation and latest trends. Proc. Inst. Mech. Eng., B: J. Eng. Manuf. **231**(9), 1670–1679 (2017). https://doi.org/10.1177/0954405415601640
2. Aboelmaged, M.G.: Predicting e-readiness at firm-level: an analysis of technological, organizational and environmental (TOE) effects on e-maintenance readiness in manufacturing firms. Int. J. Inf. Manage. **34**, 639–651 (2014). https://doi.org/10.1016/j.ijinfomgt.2014.05.002
3. Holmberg, K., Adgar, A., Arnaiz, A., Jantunen, E., Mascolo, J., Mekid, S. (eds.): E-maintenance. Springer London, London (2010). https://doi.org/10.1007/978-1-84996-205-6
4. Zhang, W., Yang, D., Wang, H.: Data-driven methods for predictive maintenance of industrial equipment: a survey. IEEE Syst. J. **13**, 2213–2227 (2019). https://doi.org/10.1109/JSYST.2019.2905565
5. Jardine, A.K.S., Lin, D., Banjevic, D.: A review on machinery diagnostics and prognostics implementing condition-based maintenance. Mech. Syst. Signal Process. **20**, 1483–1510 (2006). https://doi.org/10.1016/J.YMSSP.2005.09.012
6. Kan, M.S., Tan, A.C.C., Mathew, J.: A review on prognostic techniques for non-stationary and non-linear rotating systems. Mech. Syst. Signal Process. **62**, 1–20 (2015). https://doi.org/10.1016/j.ymssp.2015.02.016
7. Qin, S.J.: Data-driven fault detection and diagnosis for complex industrial processes. In: IFAC Proceedings Volumes, pp. 1115–1125. Elsevier (2009). https://doi.org/10.3182/20090630-4-es-2003.00184
8. Arsanjani, A.: Service-oriented modeling and architecture. IBM Dev. Work. **1**, 1–15 (2004)
9. Group, T.O.: Soa Source Book (TOGAF Series). Van Haren Publishing (2009)

10. Zimmermann, O.: Microservices tenets. Comput. Sci. Res. Dev. **32**(3–4), 301–310 (2016). https://doi.org/10.1007/s00450-016-0337-0
11. Liu, F.T., Ting, K.M., Zhou, Z.H.: Isolation forest. In: Proceedings – IEEE International Conference on Data Mining, ICDM, pp. 413–422 (2008). https://doi.org/10.1109/ICDM.2008.17
12. Liu, F.T., Ting, K.M., Zhou, Z.H.: Isolation-based anomaly detection. ACM Trans. Knowl. Discov. Data. **6**, 1–39 (2012). https://doi.org/10.1145/2133360.2133363
13. Zato: https://zato.io/. Accessed 28 May 2021
14. Ester, M., Kriegel, H.-P., Sander, J., Xu, X.: A density-based algorithm for discovering clusters in large spatial databases with noise. In: Proceedings of the 2nd International Conference on Knowledge Discovery and Data Mining, pp. 226–231 (1996)
15. Crosby, T., Iglewicz, B., Hoaglin, D.C.: How to detect and handle outliers. Technometrics **36**, 315 (1994). https://doi.org/10.2307/1269377

Check for updates

Distributing and Processing Data from the Edge. A Case Study with Ultrasound Sensor Modules

Jose-Luis Poza-Lujan[1(✉)], Pedro Uribe-Chavert[2], Juan-José Sáenz-Peñafiel[3], and Juan-Luis Posadas-Yagüe[1]

[1] Research Institute of Industrial Computing and Automatics, Universitat Politècnica de València, Camino de vera, sn 46022, Valencia, Spain
{jopolu,jposadas}@upv.es

[2] Doctoral School, Universitat Politècnica de València, Camino de vera, sn 46022, Valencia, Spain
pedurcha@doctor.upv.es

[3] Dirección de Investigación, Universidad de Cuenca, Av. 12 de Abril, 010107 Cuenca, Ecuador
juan.saenz@ucuenca.edu.ec

Abstract. Currently, the proliferation of interconnected smart devices is related to smart urban management. These devices can process sensor data in order to obtain significant information. This information can be provided to the upper layers but also can be used by the devices to take some smart actions. This article shows the change from the classic hierarchical devices and data paradigm to a paradigm based on distributed intelligent devices. The distributed model has been used to create a system architecture with Arduino-based Control Nodes interconnected by means of an I2C-bus. Each module can read the distance to each vehicle, and process this data to provide the vehicle speed and length. A case has experimented where modules share raw data and another case where modules share processed data. Results show that it is possible to reduce processing load up to 22% in the case of sharing processed information instead of raw data.

1 Introduction

Currently, intelligent systems, based on embedded devices, have been the focus of attention in smart city environments. The use of cheap and efficient microcontrollers with high connectivity features, allow devices to be integrated into almost all types of urban elements. These interconnected elements have given rise to the concept of the Internet of Things (IoT) [13]. The fact of having a large number of distributed devices implies a large amount of data to manage in order to obtain information to make some decisions. This smart distributed chain, sensor-decision-act, is aimed to provide an optimisation of system performance and to provide optimal services [1]. This optimisation has given rise to

© The Author(s), under exclusive license to Springer Nature Switzerland AG 2022
K. Matsui et al. (Eds.): DCAI 2021, LNNS 327, pp. 190–199, 2022.
https://doi.org/10.1007/978-3-030-86261-9_19

the concept of distributed intelligence, usually based on a distributed knowledge [14].

In this article, we propose the use of a distributed paradigm where the control nodes (CN) have a certain intelligence and only the relevant data is distributed, instead of using classical hierarchical models where the CN are only data providers. Based on this paradigm, the article depicts how the CN, as the basic element, works and how connecting two CN, forming an element, smart device, can improve the accuracy of the parameters measured. A smart device has been tested by means of a case study in which the smart device detects the speed and length of a vehicle. The Smart device is based on different CN, each one of them provided with an ultrasonic (US) sensor that detects distance. By changing the angle of emission of the ultrasound signal, it is possible to detect accurately the vehicle speed or length.

In the experiments, the data processing time (to obtain distances) and the processing time to generate the information (speed and length) has been measured. The results show that it is possible to reduce the overall processing time of a device, if the modules do a previous processing and provide information to other module instead of provide the raw data.

This article has been organised as follows. Next section offers a brief review of the works related to the environment of the presented system. Section 3, shows the proposed paradigm from the classical pyramid to the inverted pyramid. Section 4 presents the case study consisting of a device composed of three modules with an ultrasound sensor each one. In Sect. 5, the experiments made with two different configurations are shown. Finally, the article ends with the conclusions and possible studies to be carried out in the future research.

2 Related Work

There are many methods of detecting vehicles on roads. The most efficient use complex devices, such as cameras [11] or even drones [7]. These devices can be tempting for vandalism, in addition to not having the availability of a high processing or communicating capacity [8]. As an alternative to the previous systems, cheap solutions have been proposed. Use simple sensors, implies that the information provided by complex sensors, as cameras, must be supplied through intelligence. In [3] two classes of neural networks are considered multilayer perceptron (MLP) and convolutional neural network (CNN) to analyse the audio signal. In the case presented in the article, each CN implements a simple Long/Short Term Memory (LSTM) Neural Network. The hidden cells store the measurements and trigger the detection when the distance values changes upper the threshold.

In this context, an interesting question emerges: Is it worth distributing if an acceptable result is already obtained in a single CN? Distributing raw data loads the communications system. A high load of messages implies a high probability of error messages, long latency and variable jitter. Due to the computational resources of the CN are limited, a load increase in the communication management can lead to limiting the control algorithms. To answer the previous question it is necessary to consider the processing response time that a CN could provide. An intelligent response implies more time than a reactive response. The measurement of this response time will involve knowing the reaction times and how the action or information provided is adequate to the requirements. For example, in the experiment presented in this article, a concrete CN is used to measure the speed of a vehicle by means of an ultrasound sensor. A single sensor has a considerable error. However, if the system has multiple sensors, it is possible to improve the accuracy of the measurement by distributing the data from each CN.

3 Distributing Data for Intelligent Control

3.1 Changing the Distributed Data Model

The vision of the 'pyramid of knowledge' in intelligent control systems implies an analogy with the CN of each level (Fig. 1). A description and review of such a pyramid can be obtained in [6]. The data source (pyramid base) is the sensors that provide raw data representing physical values measured by means parameters. Raw data are processed to obtain information. Information processed produces knowledge. Finally, the knowledge is used to generate intelligence. This process is known as learning, so the initial data ends up affecting the behaviour of the system. At the pyramid low level, the data are characterised because is valid in a short time period and contains basic information. For example, a vehicle detection sensor will be able to report the vehicle presence at a specific time moment and in a certain place. As the level rises, especially if the number of sensors is increased, more useful information can be obtained.

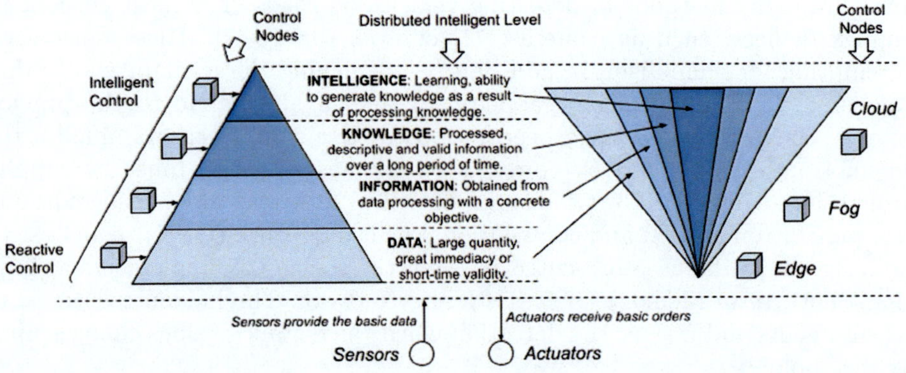

Fig. 1. Classic vision of the pyramid of knowledge and relationship with intelligent control.

As the elements of the system grow, specifically in lower layers of the classical pyramid (i.e.: a CN), a huge amount of data is available. The large number of connected elements that provide this massive data has led to the emergence of the Internet of Things (IoT) or Industry 4.0 paradigm when the IoT paradigm is applied to the socio-economic environment. Include IoT concept and Industry 4.0 paradigm into the distributed intelligent systems implies review the knowledge pyramid, such as the one proposed in [5]. Currently, there is a certain consensus to divide distributed systems in a layer close to the physical environment (Fog, Edge in the case of the hardware) and Cloud to provide massive data processing. These layers (or 'areas') have changed the way in which system architectures are designed, forcing a turn away from the hierarchical models towards highly connected horizontal models. In these new models, intelligence becomes distributed and not exclusively at the top of the classic pyramid of knowledge model (Fig. 1). Consequently, a system architecture must be able to support intelligence at the edge level and provide all the data at the cloud level. The Edge elements, as CNs, can provide some intelligent processing that helps Fog and Cloud to obtain data processed and exempt the Fog and Cloud elements to take decisions that a CN can do.

3.2 Control Node Characterisation at the Edge Level

To compare the performance of a single CN and a set of CNs working collaboratively, it is necessary to measure time spend to control processing and communication tasks. Figure 2 broadly shows the times involved in control and communication actions from a single CN. The inputs of a CN are the communications 'Comm' that receives service requests from other CN or upper elements, and the sensor data provide by the corresponding hardware elements. In turn, the outputs of the CN are the communications 'Comm' services to other Edge nodes, or system elements (cloud or fog) and the actuators 'Hw (actuators)'.

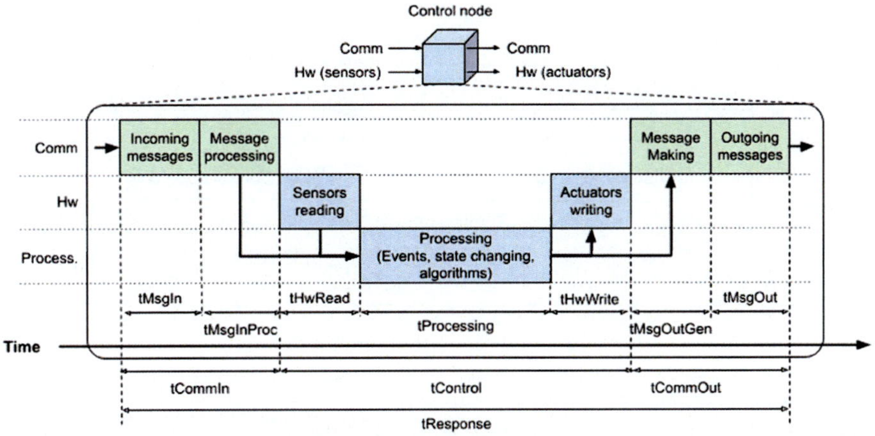

Fig. 2. Times related to communications and processing of a single control node (CN).

In this context, a CN can have different configurations. A CN without incoming or outgoing communications is an autonomous reactive node, it is not usual, but there are, for example, irrigation systems without remote monitoring or configuration. In the same way, a CN without sensors and actuators also does not make sense in the context of distributed control. A CN that only senses, is the basis of distributed sensor networks (WSN) and is widely used in human environments [9]. Control action is usually measured in terms of absolute and relative error [2]. Low error values imply effective control, even though the processing and communications consumption can arrive up to 100% of micro-controller load. However, a CN, or set of CNs, goal is obtain a low response time and low micro-controller processing load, in addition to low error values. If a control error is minimised thanks to the contribution of data, or information, from other CN, then it must be evaluated whether it is efficient to wait for the remote data or to act with a greater error.

When a CN must communicate with another, through a common communications system, then you have a distributed system. In this case, in addition to the times considered in the Fig. 2, times invested in the communication tasks between nodes should be considered as an integral part of the CN response time. All these times are shown in the Fig. 3. Times involved in this case depend on the system architecture. When nodes are on the Edge or Fog, in other words, they share a communication medium, these times are usually shorter than times involved in cloud communication.

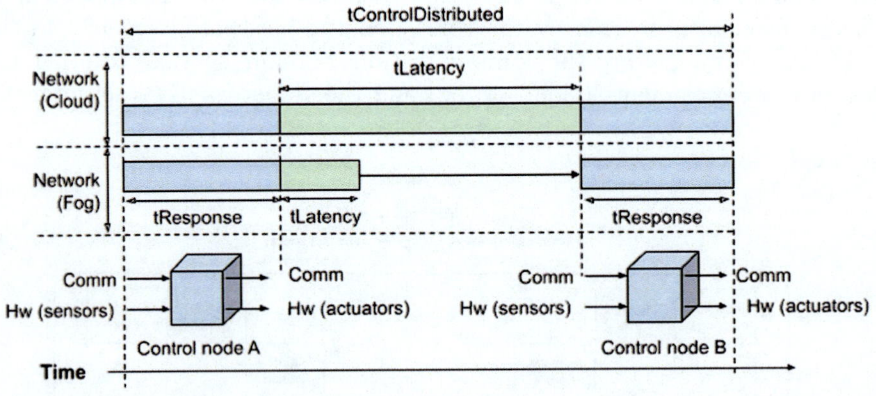

Fig. 3. Times related to communications between two different control nodes (CN) connected in the fog or the cloud.

From times outlined above and depending on the system errors, it is possible to customise from a CN to a distributed intelligent system. The following section will use these times to characterise a simple system and check which formulas can answer the question about whether it is better to act fast with a certain error, or wait a while to act, but with a minor error.

4 Case Study

The case study presented is based on a device for measuring the speed and the length of a vehicle based on the work presented in [4]. The device consists of a set of interconnected CNs. The CN has been built from the SN-SR04T 2.0. ultrasound sensor module. This sensor is waterproof and widely used in automotive for measuring the liquid tank level. The sensor has a detection range from 0.25 m to 4.5 m. This range makes it very suitable for covering both road and street lanes vehicle profiles. The resolution of the distance is 0.005 m, so it allows relevant variations of vehicle distance, so that the detection algorithm can work efficiently. The sensor sampling rate has been set to 9.6 KHz. Sensor has been connected to an Arduino Nano which communicates with the device's CNs via Inter-Integrated Circuit (I2C). This channel allows serial communication between a master and several slaves at speeds between 100 Kb/s and 3.4 Mb/s [10]. It is a channel widely used in embedded systems due to its simplicity to be managed. Left side of the Fig. 4 shows the experimental device. Depending on the orientation angle in which is set the ultrasound sensor with respect to the longitudinal axis of the road, there is a specific distance profile along the time for each detected vehicle. Figure 4 shows the three different orientations of each CN, or module: 30°, 45°, and 90° degrees. Below each module orientation is shown the signal profile obtained from distances over time.

In the case of the module oriented at 30° degrees, the number of samples that the front of the vehicle cuts the US ray is greater than in the case of the module oriented at 45° degrees. This fact allows the 30° degrees module to obtain a precise speed calculation compared with the 45° degrees module. In the case of the module oriented at 90° degrees, it can only measure the time interval in which the US sensor measures a distance less than its maximum possible measurement distance. This means that 90° degrees module can't calculate directly the vehicle speed. However, in the case that 90° degrees module could have a speed value, obtained from other CNs, the 90° module can calculate the length of the vehicle crossing in front of their US sensors.

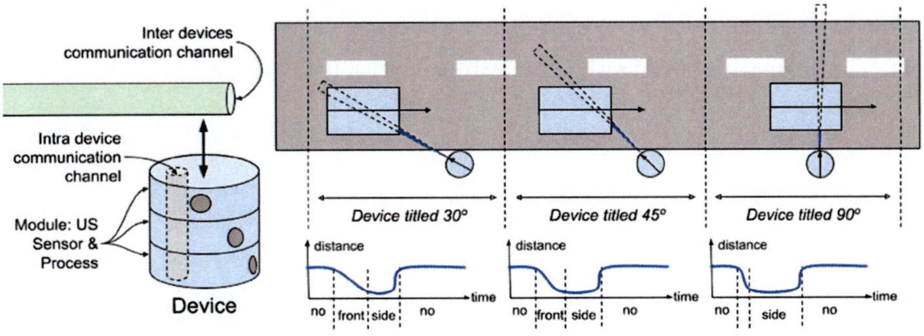

Fig. 4. Device tested (left) and method used to detect vehicle speed and vehicle length.

Broadly, each of the CN of the device shown in Fig. 4 are similar and depends on its orientation that can processing speed or distance. Once a vehicle is detected, the module proceeds to calculate the vehicle speed, detecting the change between the front and the side of the vehicle. Finally, when the sensor returns to provide the maximum distance, it is when it is considered that the transition of the vehicle has already finished. One of the most widely used neurons to store the states is LSTM consisting of forget-gate, update-gate, and output-gate. All modules perform all processing phases shown in the Fig. 5 as a cells in a LSTM Recurrent Neural Network.

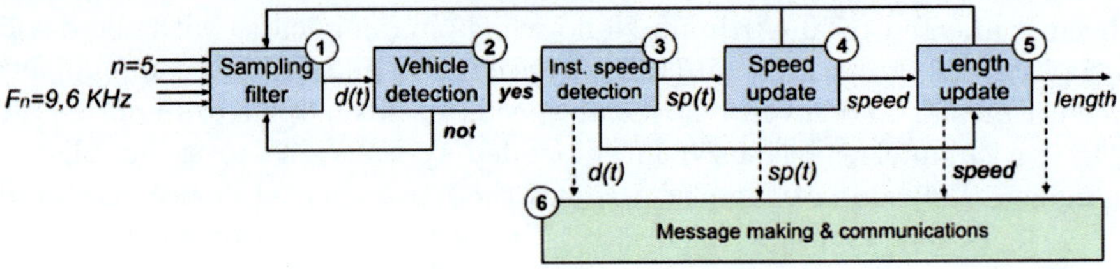

Fig. 5. Phases performed for each module (right).

The first phase (1. input-gate) consists of sampling five distances and calculating their average to obtain the distance $d(t)$. Filtering (1. input-gate) is performed discarding the samples that have different values to the sensor minimum and maximum constrains, or have a difference greater than 10% from the rest of the samples. If more than two erroneous samples are detected, the five samples are discarded. Second phase (2. forget-gate) consists in detect a vehicle. Vehicle is detected when the values $d(t) > d(t+1)$ over two consecutive operation cycles. Since a spatial and kinematic characterisation is desired, for the vehicle detection phase, the non-parametric method of frame difference will be used [12]. In this case, the recognition characteristics are the object speed and length. When an approaching vehicle has been detected, module changes to the instantaneous speed detection (3. update-gate). From the instantaneous speed detected, the phase (4. update-gate) update the speed value. Since the difference of distances detected, and time between these distances is available, the calculation of the vehicle speed is immediate. When the difference of two consecutive distances is less than 5% then it is considered that the side of the vehicle is being detected and the 'Length update' phase (5. update-gate) starts. This phase, based on the speed calculated in the previous phase, works while the vehicle is being detected, in order to determine the vehicle length. Depending on the control policy to be evaluated, it is possible to send messages (6. output-gate) in order to the processing of the phases can carry them out by a different module. This aspect, allows CN to decrease their process load, but increases the network throughput. The policy based on send data (distances) o processed data, as information about the vehicle speed and length, will be experimented in the next section.

Table 1. Results obtained from two Control Nodes (CN_) (CN_30 and CN_45) sending raw data to the third module (CN_90).

	CN_30	CN_45	CN_90	Results	CN_90
tControl (AVG)	32,34	28,97	235,73	Speed (AVG)	10,12
tControl (STD)	3,85	2,15	18,93	Speed (STD)	1,17
tResponse (AVG)	51,38	31,38	276,54	Speed (Rel.E)	5,79%
tResponse (STD)	3,18	2,18	20,49	Length (AVG)	3,76
tLatency (AVG)	1340,81	1362,25		Length (STD)	0,15
tLatency (STD)	90,90	87,34		Length (Rel.E)	1,61%

Table 2. Results obtained from the three CN processing data and sharing the information obtained. Third module (CN_90) uses the speeds calculated by the modules CN_30 and CN_45 to calculate the vehicle length and the vehicle speed.

	CN_30	CN_45	CN_90		CN_30	CN_45	CN_90
tControl (AVG)	45,51	49,65	113,04	Speed (AVG)	10,55	12,92	10,08
tControl (STD)	5,34	1,44	1,06	Speed (STD)	1,01	1,18	0,97
tResponse (AVG)	105,16	56,67	132,22	Speed (Rel.E)	5,79%	17,77%	4,12%
tResponse (STD)	1,77	4,77	12,06	Length (AVG)	NA	NA	3,71
tLatency (AVG)	1339,41	1362,65	NA	Length (STD)	NA	NA	0,09
tLatency (STD)	92,30	85,94	NA	Length (Rel.E)	NA	NA	0,78%

5 Experiments and Results

The experiment performed consists of the characterisation of a single module, using three different CNs with a US sensor oriented with different angles: 30°, 45° and 90°, as described in the Fig. 4. For the experiments, a vehicle with a real length of 3.7 m was used. The vehicle has been passed through the measuring module ten times for angles of 30°, 45° and 90° degrees. For each of the angles, the vehicle speed has been 10 m/s (36 km/h). Measured times are all in milliseconds (ms).

The aim of the device is to obtain an accurate measure of the vehicle speed and length. To achieve this accuracy the CNs must collaborate among them. The modules configured with 30° and 45° degrees can obtain vehicle speed with certain accuracy, and a module configured with 90° degrees can obtain the vehicle length using the speed obtained from the previous angled CNs. Two different cases have been tested. The first case represents central processing and the second case represents a distributed processing. In the first case, the 30° and 45° modules transmit the raw data of the measured distances to the 90° module. This case corresponds to a hierarchical model in which the angled CNs are dedicated to data collection, while the 90° node is dedicated to both collecting data and calculating the resulting speed and length. Results are shown in the Table 1.

In the second case, the 30° and 45° degrees CNs calculate the speed and length average, in conjunction with the standard deviation, and this information is sent to the 90° degree module. Results are shown in the Table 2.

The results obtained show how the speed and length calculated by the CN_90 module have a similar accuracy that the previous case. This result is expected, but isn't the aim of the experiment. We need to compare the full time involved in both cases If we calculate the overall time that the device dedicates to the process, we obtain 315.04 ms in the first case, but in the second case, the total process time is 208.00 ms. Indeed, if we consider the response time, thus it is the sum of communications and process times, the results are 359.3 ms in the first case, and 294.05 ms in the second one. This result means that a distributed processing saves around 22% of processing and communications time. All the results show how distributing the processing between CN in the device decreases the overall processing time. Although the latency time has no significant changes, since modules work in an I2C network, there is a big amount of data transferred in the first case. This is because, for each message with the instantaneous speed transmitted, five messages must be transmitted with the measured distances. This difference is even greater in the case of transmitting only the final speed or the final transmitted length. Reduce messages and processing data before send it has relevant implications, especially for power consumption due to the processing time and network load.

6 Conclusions

This article has presented a paradigm that allows modules to share data and also share processed information. Based on the paradigm, a module called Control Node (CN) has been presented and characterised. A CN has been implemented as part of a device that obtains the speed and the length of vehicles by means of ultrasonic sensors. It is possible to prove how the modules collaboration at the edge level, provides an improvement in the quality of the information, measured in terms of the relative error. As experiments have been proven, process data close to the CN reduce the overall time dedicated to processing in global terms. These results open the door to future experiments where, in addition to sharing information within a device, information is shared through the fog between devices. The overall power consumption of the system can be reduced. This reduction implies that study how processing data close to the edge level is a good starting point to new experiments. Aspects like message transmission policy based on the relative error to reduce non-relevant information can be used to optimise the system performance. As a future research line, we propose to implement CN as cells of a distributed Neural Network that, dynamically, can select which kind of information suits to be distributed.

Acknowledgements. Work supported by the Spanish Science and Innovation Ministry MICINN: CICYT project PRECON-I4: "Predictable and dependable computer systems for Industry 4.0" TIN2017-86520-C3-1-R.

References

1. Amurrio, A., Azketa, E., Gutierrez, J.J., Aldea, M., Parra, J.: A review on optimization techniques for the deployment and scheduling of distributed real-time systems. Revista Iberoamericana de Automática e Informática Industrial **16**(3), 249–263 (2019)
2. D'Andrea, R., Dullerud, G.E.: Distributed control design for spatially interconnected systems. IEEE Trans. Autom. Control **48**(9), 1478–1495 (2003)
3. Golovnin, O., Privalov, A., Stolbova, A., Ivaschenko, A.: Audio-based vehicle detection implementing artificial intelligence. In: Dolinina, O. et al. (eds.) Recent Research in Control Engineering and Decision Making. ICIT 2020. Studies in Systems, Decision and Control, vol. 337, pp. 627–638. Springer, Cham (2021). https://doi.org/10.1007/978-3-030-65283-8_51
4. Bel, A.H.: Dispositive modular configurable para la detección de vehículos, y viandantes, y con soporte a la iluminación de la vía e información de tráfico. Master's thesis, DISCA, UPV, Valencia (2020)
5. Jennex, M.E.: Big data, the internet of things, and the revised knowledge pyramid. ACM SIGMIS Database **48**(4), 69–79 (2017)
6. Körner, M.-F., et al.: Extending the automation pyramid for industrial demand response. Procedia CIRP **81**, 998–1003 (2019)
7. Li, W., Li, H., Wu, Q., Chen, K., Ngan, K.N.: Simultaneously detecting and counting dense vehicles from drone images. IEEE Trans. Ind. Electr. **66**(12), 9651–9662 (2019)
8. Dominguez, J.M.L., Sanguino, T.J.M.: Review on v2x, i2x, and p2x communications and their applications: a comprehensive analysis over time. Sensors **19**(12), 2756 (2019)
9. Vicente, E., Merchán, M., Francisco, I., Pina, B., Ricardo Núñ, J., Alvarez, N.: Network of multi-hop wireless sensors for low cost and extended area home automation systems. RIAI-Rev. Iberoam. Autom. Inform. Ind. (2020)
10. Semiconductors, P.: The i2c-bus specification. Philips Semicond. **9397**(750), 00954 (2000)
11. Sun, J., Bebis, G., Miller, R.: On-road vehicle detection using optical sensors: a review. In: Proceedings of the 7th International IEEE Conference on Intelligent Transportation Systems (IEEE Cat. No. 04TH8749), pp. 585–590. IEEE (2004)
12. Weng, M., Huang, G., Da, X.: A new interframe difference algorithm for moving target detection. In: 2010 3rd International Congress on Image and Signal Processing, vol.1, pp. 285–289. IEEE (2010)
13. Xia, F., Yang, F.T., Wang, L., Vinel, A.: Internet of things. Int. J. Commun. Syst. **25**(9), 1101 (2012)
14. Zare, R.N.: Knowledge and distributed intelligence. Science **275**(5303), 1047–1048 (1997)

Check for updates

Bike-Sharing Docking Stations Identification Using Clustering Methods in Lisbon City

Tiago Fontes[1,2](✉), Miguel Arantes[2], P. V. Figueiredo[2], and Paulo Novais[1]

[1] Universidade do Minho, Braga, Portugal
tiago.fontes@ceiia.com, pjon@di.uminho.pt
[2] CEiiA, Matosinhos, Portugal
{antonio.asilva,paulo.figueiredo}@ceiia.com

Abstract. Urban clean mobility has enormous impacts on environmental, economic and social levels, promoting important eco-friendly means of sustainable transportation. Soft mobility (specially bike-sharing services) plays a crucial role in these initiatives since it provides an alternative for hydrocarbon fuel vehicles inside the cities. However, choosing the best location to install soft mobility docks can be a difficult task since many variables should be considered (e.g. proximity to bike paths, points of interest, transportation access hubs, schools, etc.).

On the other hand, mobile data from personal cellphones can provide critical information regarding demographic rate, traffic trajectories, origin/destination points, etc., which can aid in the installation of soft mobility platforms.

This paper presents a decision support system to study both existent and new bike-sharing docking stations, using mobile data and clustering techniques for three Lisbon council parishes: *Beato*, *Marvila* and *Parque das Nações*.

1 Introduction

In the last decades, overpopulation growth has created several challenges in urban centres related to climate changes. Moreover, this growth has raised new challenges in environmental, economic and social levels [1]. On the other side, both European Commission and the United Nations have been approaching Climate change issues as proprietary actions to be developed.

In Portugal, cities like Aveiro and Lisbon have been pioneers in promoting new sustainable technologies, following the digitisation process across these cities.

Hence, soft mobility plays an important role in sustainable technologies in several platforms, being one of them the shared Bicycle Systems (SBP). However, implementing soft mobility solutions entails high costs [2].

Portugal has also been promoting several initiatives focused on green transportation to reduce greenhouse emission gases within the cities. These initiatives

© The Author(s), under exclusive license to Springer Nature Switzerland AG 2022
K. Matsui et al. (Eds.): DCAI 2021, LNNS 327, pp. 200–209, 2022.
https://doi.org/10.1007/978-3-030-86261-9_20

are driven by the related United Nations Sustainable Development Goals (objectives: 11 - Sustainable Cities and Communities; 13 - Climate Action).

The term Soft mobility includes any non-motorised transportation system (human-powered mobility) [3]. Thus, soft mobility is related to pedestrian, bicycle, roller skate and skateboard transportation, which can be defined as a particular form of sustainable mobility capable of optimising urban livability (known as mobility models).

A mobility model details moving patterns for a set of users, showing location, velocity, acceleration, etc., over time. Being aware of all soft mobility advantages (and benefits to society), bicycles are one of the most common expected drivers that may potentially change city carbon dioxide emission.

Since the beginning of the century, bicycle-sharing schemes have been rapidly growing in several earth regions, being considered as an indicator of the attractiveness of such systems and their adaptability to changes according to different situations [4].

Although the benefits to society, some major obstacles are number and location definitions for new docking stations. Defining these outputs are not an easy task since several parameters must be considered. Moreover, these parameters can change through time, including the demographic rate, tourism interest points, transportation access hubs, bicycle paths, time of the year, year's season, etc., for each parish council.

In this context, it is crucial to develop an algorithm capable of forecasting docking stations according to the previous parameters. Extra complexity levels can be added to the study, such as socioeconomic characteristics of a given population (e.g. monthly income, development index, availability of public transportation, education, climatology, etc.) for an even preciser prediction. However, these extra parameters are not used in this study.

1.1 Lisbon Bicycles

Lisbon has one of the most common and well-known Bike-sharing Schemes (BSS) in the world. Created in 2017, the public bike-sharing service was created in the city centre (named here as *Lisbon Bicycles*). Municipal Mobility and Parking Company have operated the system after a pilot phase in *Parque das Nações*.

At the end of 2018, the mark of one million trips was reached [10]. Nowadays, more than three and half million trips were performed. Hence, the noticeable and continuous increase in popularity of the service has allowed rapid growth of docking stations, bicycles etc. For this year, the number of docking stations and bicycles are expected to rise, following the intended investments predicted for bicycle paths for the upcoming years [11,12].

In 2017 the investment was performed for *Parque das Nações*. Nowadays, the service is expected to be available in two more council parishes (*Marvila* and *Beato*).

These new locations do have several tourism access points, having also public transportation access hubs (subway, train, etc.), and new bike paths under developement. Figure 1 presents *Parque das Nações*, *Marvila* and *Beato* parishes.

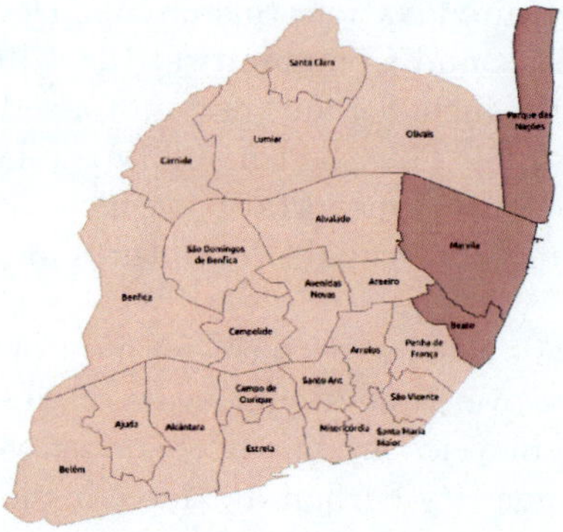

Fig. 1. Map of Area of Interest (AoI) - *Beato*, *Marvila* and *Parque das Nações*

1.2 Related Work

Bike-sharing systems are in the extensive and continuous investigation. Hence, security plays an important part for all users [17]. This work also presents bike-sharing usage as a replacement for short trips and first/last mile usage, correlating data with other existing mobility services. There is a direct relationship between bicycle usage and high-quality bicycle paths, which means that bicycle research is essential when planning bicycle-sharing docking stations [18].

Several researchers have used cluster techniques for urban mobility in the last years [6]. Using clustering algorithms like K-medoids [7], spectral clustering algorithm [8] and the agglomerative clustering [9], it was possible to find trajectories with similar location patterns [5].

In [19], a clustering process was performed in order to categorise all the stations into several groups. In addition, this experience demonstrated temporal patterns of rush-hour clusters (weekdays), gathering groups of bicycle sharing docking stations [20].

Considering the state of the art analysed, clustering applications (k-means) were verified to be a significant methodology for mapping bicycle-sharing services, which was used for this study.

2 Methodology

For the development of this work, two main subsections were developed. First, the data is verified, explaining both origin, parameters and constraints. Then, the process is fully detailed, explaining the used methodology and algorithms.

It is known that there are already existent in-situ mobility sharing services (*Parque das Nações*, *Beato* and *Marvila*) with several docking stations. Therefore, for the development of this study, these docking services were considered and used as a baseline.

2.1 Data

Telecommunication operators over the whole world generate high data volumes. Each mobile device can act as a tracking device, providing vital information regarding where each device went, which can be used to analyse patterns such as location, PoI, people clusters, specific events, etc. However, this data is susceptible since it deals with personal information and shall not compromise personal human rights.

There are several telecommunication operators in Portugal, in which all of them are subjected to General Data Protection Regulation (GDPR) [13] and also ethical and legacy principles. In January 2020, one operator reached over 70000 entries.

The representation of high people clustering is performed in the vector form (as a polygon) over a GIS (Geographic Information System), also named as Bin or S2 cell [14].

Although the large data set, this information also presents several constraints. Considering the scope of this work, the following constraints were considered:

1. A device might be detected and considered in many entries of the dataset
2. Geographical definition of the area (bin) might be affected by GPS errors and limitations
3. Velocity is the only available attribute assumed for mobility purposes (e.g. walking, running, cycling, stopping, etc.).

Following the GDPR guidelines, the data from each entry can be summarised in Table 1, presented below.

Figure 2a and b presents and example of two mobile data entries from *Parque das Nações*.

Table 1. Data structure and meaning of each attribute

Column name	Data type	Meaning
Identifier	Integer	Unique identifier for each record in dataset. It has no correlation with devices identification
Day	Date	Date when device was detected in the given area
Hour	Integer	Corresponds for all the available time frames during the day. Temporal granularity for all records
Speed	Float	Means the average speed of the devices detected in a given area
Number of devices	Integer	Indicates the number of aggregated devices for specified entry in the dataset
Bin	MultiPoint	S2 Cell defined by list of points given in ESPG:4326. Defines an area similar to a square of 50×50 m

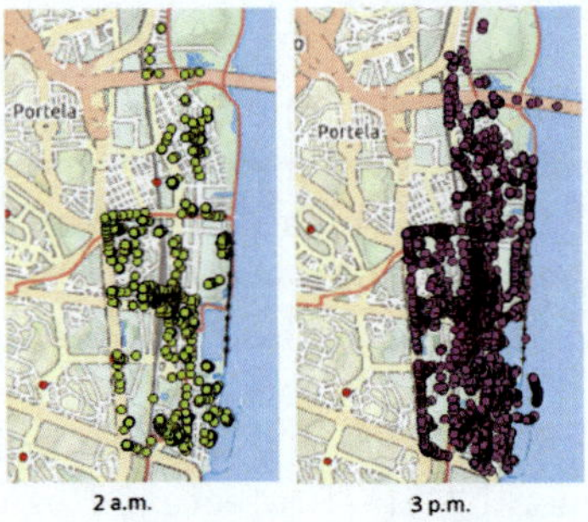

Fig. 2. Data visualisation in *Parque das Nações*: 2a) 2 a.m.; 2b) 3 p.m. of a weekday (January 2020). During the night most residential areas, neighbourhoods are highlighted, during day more devices can be observed, alongside as main avenues and roadways, tourism points, etc.

2.2 Process

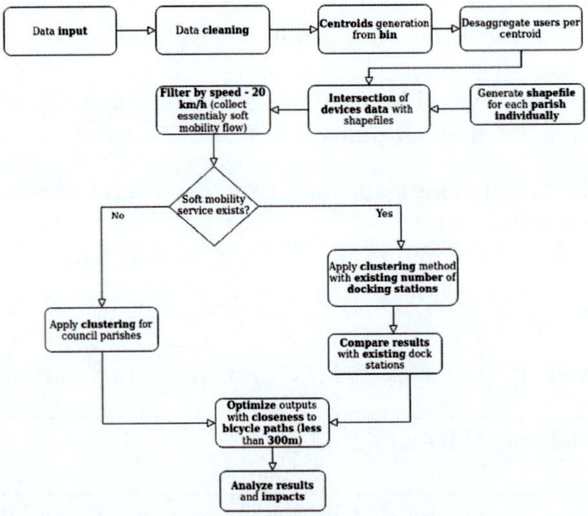

Fig. 3. Methodology applied

As stated previously, mobile data was received by a Portuguese telecommunication operator. In this context, the first step consisted of developing a methodology pipeline, which started with acquiring a data-gathering mechanism, then the data was cleaned and analysed.

Some parameters were kept to be used during the analysis and cleaning process, and others were removed (e.g. time frame availability of the data, external factor, public holiday, etc., since they would affect mobility patterns).

Due to privacy policies and according to GDPR, it was not possible to work with exact geographic coordinates. Therefore, it was necessary to consider each *bin* as the geographical characterisation of each record (area). Using area parameter for each record, it was verified that it added extreme complexity to the analysis since more than one point was being considered for each record, resulting in a *MultiPoint* structure type. This point led to a necessity for complexity reduction (centroid calculation). The centroid calculation was performed aided by GIS software (finding the centre of each polygon for sets of individual points).

After the previous point, it was necessary to create an equitable distribution of our device points (since they have been aggregated), then an unlist operation was performed. In this sense, each entry was repeated throughout the dataset N times, being N the number of devices considered in that capture. Figure 3 presents the methodology used in this work.

Due to different geographical asymmetric characteristics in our AoI, (Sect. 1), and considering the existence of soft mobility solutions just for one of the studied parishes[1], a comparison analysis was necessary for algorithm validation. The algorithm is expected to perform similar results in BSS stations location.

Aiming the individual study in each parish, a geoprocessing operation was performed—this procedure allowed to delimit the geographic distribution of all docking points.

For the development of the algorithm, an average velocity of 20 km/h was used [15,16], independent of the user gender, physical condition, weather conditions, among others.

For this work **K-means** [21] algorithm was used. Hence, the first step when running k-means consisted of defining the number of clusters that should match the same number already installed of BSS docks. So far there are fourteen docking stations in *Parque das Nações*. The same number of docking stations was considered when running k-means. In the development of this work, the **Elbow Method** was used for the K-means algorithm.

In order to describe in simple way the optimisation process, Algorithm 1 presents a pseudo-code explanation regarding this process.

Algorithm 1: Optimization for bike-sharing docking stations

```
Result: List of docking stations nearby bike paths
Apply K-Means algorithm on specific council parish;
Define threshold for otimization (in meters);
For each centroid(c) output from k-Means do    For each segment(seg)
 of bike-path do      Project c in seg, output is np, following nearest
 point definition;
     Calculate distance(d) from np and c;
  If the minimum distance from d to np is lower than threshold;
     Adopt the point in segment as optimal point (located in bike-path);
  EndDo;
EndDo;
Print optimised points;
```

[1] *Parque das Nações.*

3 Discussion and Results

Table 2 presents the Sum of Squared Errors for best K for all council parishes.

Table 2. SSE - Sum of Squared Error for best K, following *Elbow method. Marvila* is a larger parish than *Beato* so is the error. *Parque das Nações* has the same number of docking stations.

Council parish name	Number of clusters	SSE - Sum of Squared Errors
Parque das Nações	14	0.853283
Marvila	8	1.662275
Beato	5	0.462912

3.1 Parque das Nações

After running the analysis for *Parque das Nações*, results were observed (Fig. 4).

The map was divided into three sections, showing the differences between existing docking stations (*Lisbon Bicycles*) and used methodology. Analysing the data in detail, it is possible to verify that there are fourteen points for both cases.

From a comprehensive perspective, most yellow points are next to dark blue points. In those dots, the algorithm has worked as planned (having a difference under 200 m). Therefore, only the dark blue dots away two hundred meters for the yellow dots will be analysed in detail.

In the first section, there are two separate yellow dots in which the next dark blue dot is a little south. The blue, dark blue dot is next to a picnic garden and a

Fig. 4. Bike-sharing docking stations in *Parque das Nações*: Existing docks in-site are represented by yellow dots; red dots represent centroid clusters; dark blue dots represent the optimised locations for k-means outputs (docking stations over bike paths). Subway has the pink colour polygon while the purple polygon regards the train station. The map was divided in three segments.

significant collegiate, a nursery and next to a driving school. These PoI led to the difference between them. In the second section, there is a dark blue dot isolated (upper east). This dot has several PoI, such as juvenile garden, is next to Tagus river and has several main bike paths, which also justifies the need for a docking station here. In the third section, three dark blue dots do not have any yellow dot nearby. The first dot (upper dot in the third section) was implemented since there is one nearby subway station (pink colour polygon *Moscavide*). The second point is nearby one of the most important train stations in Portugal *Gare do Oriente*, which provides transportation for thousands of people daily. As for the last dot (lower left side of the section), the algorithm assumes the subsequent bike paths in development, being previously considered for the analysis.

3.2 Beato and Marvila

Beato is the smallest council parish in analysis in this study. Nevertheless, it still provides some interesting/crucial urban mobility flows, and it is possible to identify exciting spots for bicycle dock stations.

Hence, considering the distribution of SSE, five centrois was used for *Beato* and eight for *Marvila* (see Table 2). Figure 5 presents the location of the docking stations in both council parishes. In this map, the geographical distribution of new docking stations is represented in dark blue dots. Regarding Beato, Fig. 5a, and considering *Olaias* subway station, having the *Olaias Plaza* shopping centre in the same location. As it can be seen, both nearby Tagus river dots are on the bike paths, then two dots are next to schools, and the final dot is very close to both the shopping centre and subway station.

Figure 5b illustrates the distribution of dock stations for *Marvila*. By observing subway stations and the dock stations locations suggested, we can identify a correlation between subway stations and output, highlighting the importance of these soft mobility solutions for last-mile trips and nearby other mobility infrastructures. The other dots are close to PoI (schools and green spaces, bicycle paths, etc.).

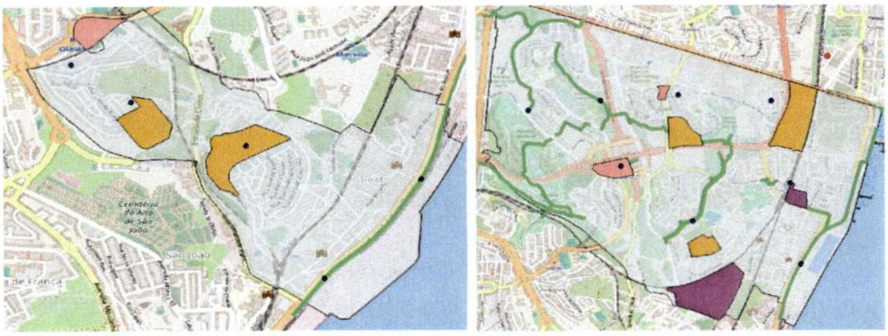

Fig. 5. Bike-sharing docking stations prediction: 5a) *Beato* and 5b) *Marvila*. Dark blue dots represent the optimised locations for k-means outputs (docking stations over bike paths). Subway has the pink colour polygon while the purple polygon regards the train station. Schools are coloured yellow and bike paths green.

4 Conclusion

Urban mobility represents an issue of extreme importance for climate change purposes alongside with United Nations sustainable goals. Hence, it is important to use sustainable means of transportation, increasing the efficiency of existing systems already developed.

The main objective of this paper was to provide a beneficial data correlation from bicycle docking stations alongside GSM data. This data was later combined with GIS parameters, such as current mobility services, (e.g. metro and train stations, bike paths, etc.) from a sustainable perspective.

Soft mobility can still be used as an active Eco-friendly city solution when compared with other transportation means. A decision support system based on clustering techniques was used, having the same number of docks installed on-site. This system was proved to be a powerful tool in developing new docking stations and soft mobility.

In a first step, the data was acquired and treated, and then the presented model was used for the parishes of *Parque das Nações*, *Marvila* and *Beato*. This methodology was possible to forecast optimal locations for bike-sharing dock systems, which was verified to be similar to those verified on-site, validating this study.

Since the European Commission has already identified Urban Mobility as a priority issue for the new decade (accounting for 40% of all CO^2 emissions) in the Clean Transport and Urban Transport section, the development of new and green solutions must be adopted for facing problems such as transportation and traffic jams.

Acknowledgement. This work has been supported by FCT – Fundação para a Ciência e Tecnologia within the R&D Units Project Scope: UIDB/00319/2020.

References

1. Singh, R.P., Singh, A., Srivastava, V.: Environmental issues surrounding human overpopulation. Information Science Reference (2017)
2. De Maio, P.: Bike-sharing: history, impacts, models of provision, and future. J. Public Transp. **12**, 3 (2009)
3. La Rocca, R.: Soft mobility and urban transformation. TeMA J. Land Use Mob. Environ. **2** (2010)
4. Midgley, P.: Bicycle-sharing schemes: enhancing sustainable mobility in urban areas (2011)
5. Ben-Gal, I., Weinstock, S., Singer, G., Bambos, N.: Clustering users by their mobility behavioral patterns. ACM Trans. Knowl. Disc. Data **13**, 1–28 (2019). https://doi.org/10.1145/3322126
6. Lee, M., McKenzie, G., Aghi, R.: Exploratory cluster analysis of urban mobility patterns to identify neighborhood boundaries (2017)
7. Kaufman, L., Rousseeuw, P.: Clustering by means of medoids. In: Dodge, Y. (ed.) Statistical Data Analysis Based on the L 1-Norm and Related Methods, pp. 405–416. North-Holland, Amsterdam (1987)

8. Ng, A.Y., Jordan, M.I., Weiss, Y.: On spectral clustering: analysis and an algorithm. Adv. Neural. Inf. Process. Syst. **2**, 849–856 (2002)
9. Day, W.H.E., Edelsbrunner, H.: Efficient algorithms for agglomerative hierarchical clustering methods. J. Classif. **24**, 7–24 (1984)
10. Lisbon Bicycles 2018 statistics. https://www.emel.pt/pt/noticias/bicicletas-gira-ja-rolaram-1-milhao-de-viagens-2-2/. Accessed 16 May 2021
11. Increase number of *Lisbon Bicycles* 2021. https://www.sabado.pt/portugal/detalhe/emel-estima-duplicar-numero-de-bicicletas-gira-em-lisboa-ate-2021. Accessed 14 May 2021
12. More 700 bicycles in Lisbon until March 2021. https://observador.pt/2020/12/30/lisboa-com-mais-700-bicicletas-eletricas-ate-final-marco-de-2021/. Accessed 12 May 2021
13. General Data Protection Regulation (GDPR). https://gdpr-info.eu/. Accessed 5 May 2021
14. S2 Gemetry. https://s2geometry.io/. Accessed 2 May 2021
15. What's the average speed of a beginner cyclist? https://www.roadbikerider.com/whats-the-average-speed-of-a-beginner-cyclist/. Accessed 1 May 2021
16. Jensen, P., Rouquier, J., Ovtracht, N., Robardet, C.: Characterizing the speed and paths of shared bicycle use in Lyon. Transp. Res. Part D Transp. Environ. **8**, 522–524 (2010)
17. Wang, J., Huang, J., Dunford, M.: Rethinking the utility of public bicycles: the development and challenges of station-less bike sharing in China (2019)
18. NACTO: Bike Share Equity Practitioners' Paper #3 July 2016. Equitable Bike Share Means Building Better Places For People to Ride
19. Feng, Y., Affonso, R.C., Marc, Z.: Analysis of bike sharing system by clustering: the Vélib' case. In: IFAC 2017, Toulouse, France, July 2017
20. Ma, X., Cao, R., Jin, Y.: Spatiotemporal clustering analysis of bicycle sharing system with data mining approach. Information **10**, 163 (2019)
21. Keller, J.M., Gray, M.R., Givens, J.A.: A fuzzy k-nearest neighbor algorithm. IEEE Trans. Syst. Man Cybern. **4**, 580–585 (1985)

Check for updates

Development of Mobile Device-Based Speech Enhancement System Using Lip-Reading

Fumiaki Eguchi[1], Kenji Matsui[1](✉), Yoshihisa Nakatoh[2], Yumiko O. Kato[3], Alberto Rivas[4], and Juan Manuel Corchado[4]

[1] Osaka Institute of Technology, Osaka, Japan
kenji.matsui@oit.ac.jp
[2] Kyushu Institute of Technology, Kitakyushu, Japan
[3] St. Marianna University School of Medicine, Kawasaki, Japan
[4] BISITE Digital Innovation Hub, University of Salamanca, Salamanca, Spain

Abstract. We have been developing a speech enhancement device for laryngectomees. Our approach is to use a lip-reading technology to be able to recognize Japanese words from lip images and generate speech outputs using mobile devices. The target words are translated into registered 36 viseme sequences, and converted into VAE (Variational Auto Encoder) feature parameters. Then the corresponding words are recognized using CNN-based model. Previously the PC-based experimental prototype was tested with 20 Japanese words and a well-trained single subject. From the result, we confirmed 65% recognition accuracy, and 100% including 1st and 2nd candidates from that test. In this paper, a couple of recognition performance improvement methods were investigated with larger vocabulary and multiple subjects. After adjusting the speech rate and the mouth movement, we obtained about 60% word recognition accuracy including the 1st through 6th candidates with inexperienced users. Also, we developed a mobile device based prototype and conducted the preliminary recognition experiment with 20 words by a well-trained single subject, and 95% accuracy was obtained including the 1st through 6th candidates, which was almost equivalent to the PC-based system.

Keywords: Lip-reading · Laryngectomy · Viseme

1 Introduction

There are several options for laryngectomees to be able to communicate. The electrolarynx is easy to use, however, the output speech sounds monotonous and has no intonation because it just uses a simple vibration mechanism [2]. Also, it is difficult to generate consonants. Secondly, it is far from normal appearance, so that some users do not want to use it actively. Alternatively, esophageal speech does not require any special equipment, but requires speakers to swallow air into the esophagus. In fact, many elderly laryngectomees face difficulty in mastering the esophageal speech. Also it is difficult to keep using esophageal speech. Resulting the waning strength, the output speech becomes weak, and low pitch.

© The Author(s), under exclusive license to Springer Nature Switzerland AG 2022
K. Matsui et al. (Eds.): DCAI 2021, LNNS 327, pp. 210–220, 2022.
https://doi.org/10.1007/978-3-030-86261-9_21

We have done various speech enhancement approach to improve the sound quality of esophageal speech and also improve the usability of electrolarynx. First, we prototyped a real-time speech analysis-synthesis device, which extracts the voice source signal and formant parameters to be able to re-construct the input speech [1]. We also developed a small and light-weight electrolarynx, which you can put on your neck without holding by hand. However, according to the user's evaluation, none of those approach was satisfactory. Based on our user needs survey with 121 laryngectomees, we focused on the three strong demands [2, 3].

1) "Use of existing devices": Mobile phone, especially smartphone is becoming very popular and it has a lot of computational power. Therefore, we might be able to use it as the central unit of speech enhancement system.
2) "Ordinary-looking": No one thinks it is strange even if you are talking to your mobile phone. We plan to develop a system which can recognize user's lip motion and generate the corresponding synthesized speech.
3) "Easy to use": By combining lip-reading and speech synthesis, people could communicate without using either electrolarynx or esophageal speech.

In response to such feedback from users, we have been developing technology that supports conversation using lip-reading and speech synthesis [4–7].

By developing an application software that can recognize people's lip motion and interact with speech synthesis output, it is possible to provide a communication support system that can be used at any time simply by downloading the software. LipNet is a speaker independent, sentence level lip-reading system, which is considered to be state-of-the-art technology in lip-reading research [13]. However, good recognition performance has not been seen so far in the case of Japanese. In terms of our target system, it requires speaker-dependent mobile phone application with quick user adaptation. So far, we have developed a word recognition algorithm based on 36 viseme units. First, viseme sequence images are converted into VAE feature vector sequences. Then a CNN word recognizer is used to recognize those words. Although the system is a speaker dependent, small to mid-size vocabulary recognition system, it is easy to add the vocabulary words by just typing the kana-characters and re-train CNN. In this paper, we analyze the performance difference depend on the type of users, vocabulary size. Also, we develop applications on mobile terminals, and design user interfaces [8–10].

2 Lip-Reading Method Using VAE

Figure 1 shows the processing flow of the lip image extraction. It captures the input image at 30 fps. First, a face image is converted into a HOG feature. The gradient direction and gradient intensity of the cell brightness in the image is calculated. After that, a histogram is created and normalization is performed in each block area, so that the system will be robust against geometric transformation and fluctuations in lighting. The face detection library uses HOG feature with SVM, then, Gradient Boosting Degnition Tree (GBDT) was used to detect the lip region.

Histogram normalization is performed on the area image to obtain 64 × 64 pixels [11, 12]. Figure 2 shows lip images of the preprocess output. Those images are Japanese five vowels and closure X.

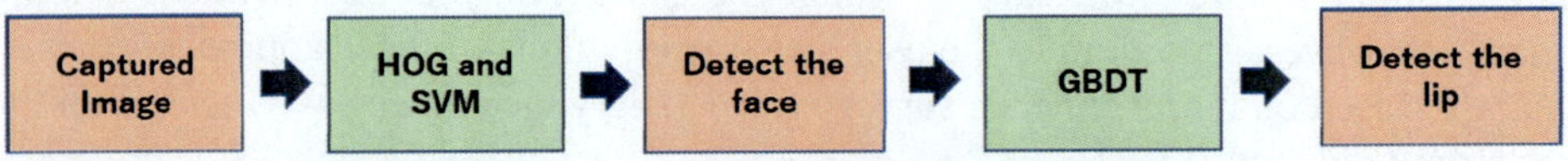

Fig. 1. A block diagram of lip region image extraction. HOG: histogram of oriented gradients, SVM: support vector machine, GBDT: gradient boosting decision tree

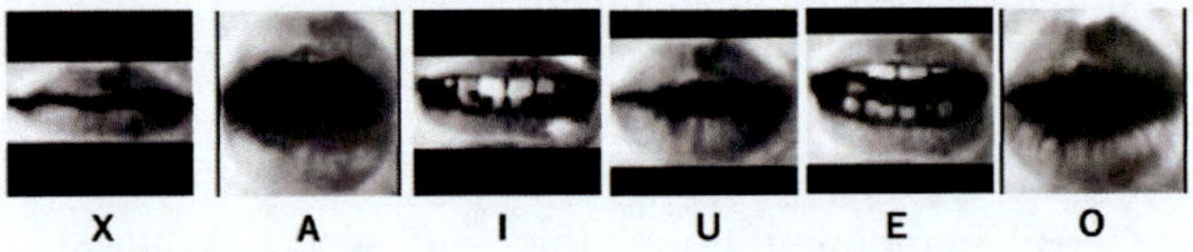

Fig. 2. Preprocessed viseme images

Table 1 shows the 36 syllable patterns of visemes including closure X. Those viseme movies were captured and used for the training. In the case of single phoneme, while capturing the image, the speaker needs to move the face slightly from side to side, up and down, and diagonally while uttering the listed vowels. In the case of syllables, recording is started while speaking the first vowel and the recording is finished while uttering the second vowel.

Table 1. 36 patterns of visemes consist of five Japanese vowels and closure (X)

X	A	I	U	E	O
XA	AX	IX	UX	EX	OX
XI	AI	IA	UA	EA	OA
XU	AU	IU	UI	EI	OI
XE	AE	IE	UE	EU	OU
XO	AO	IO	UO	EO	OE

Variational Auto Encoder was used to extract the features of viseme images. Figure 3 shows the configuration of VAE. The VAE encoder is a model with multiple convolution layers, and takes an image as an input to obtain the latent representation space z. Normally z is a standard normal distribution and specified by average vector μ and variance vector σ. The encoder was trained by setting the number of dimensions of the latent variable Z to 3 dimensions. 36 viseme images were recorded from one male speaker. VAE was trained using those images. For each of those 36 image data, five consecutive images were extracted around the highest frame difference point. Then, using those processed data, the feature vector sequences were generated. Figure 4 shows the generation of the vector sequences. The largest part of the optical flow is the center of the sequence. A sequence of viseme corresponding to hiragana was created.

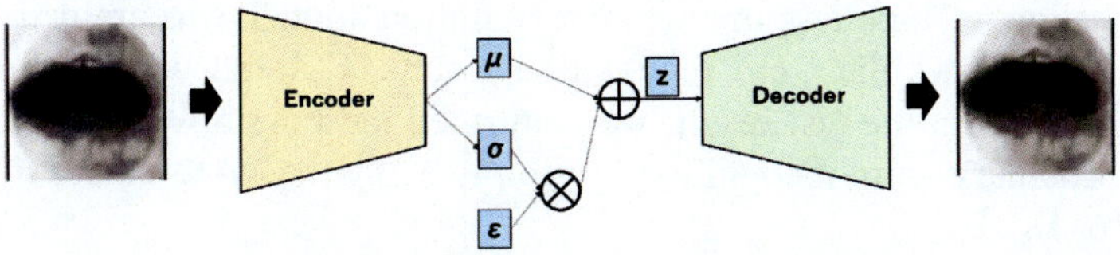

Fig. 3. Configuration of the VAE model

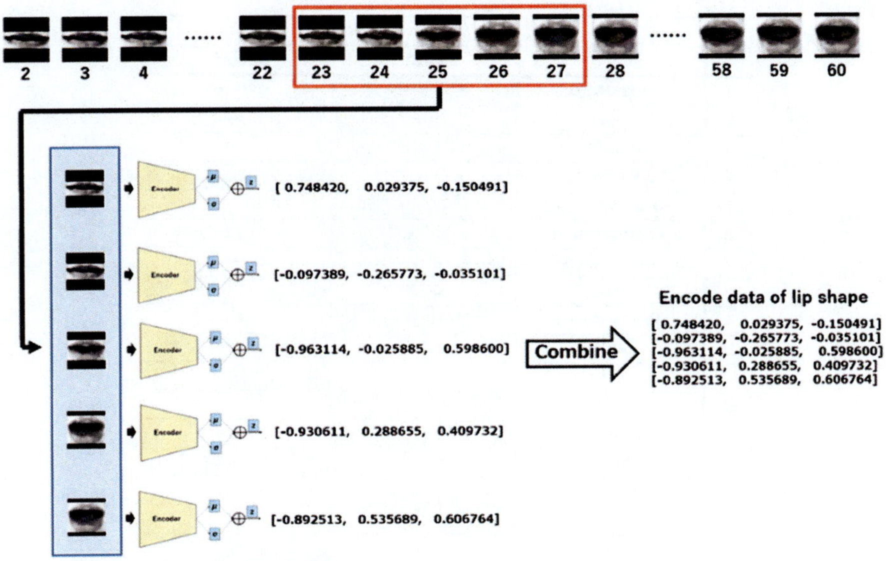

Fig. 4. Generation of the feature vector sequence

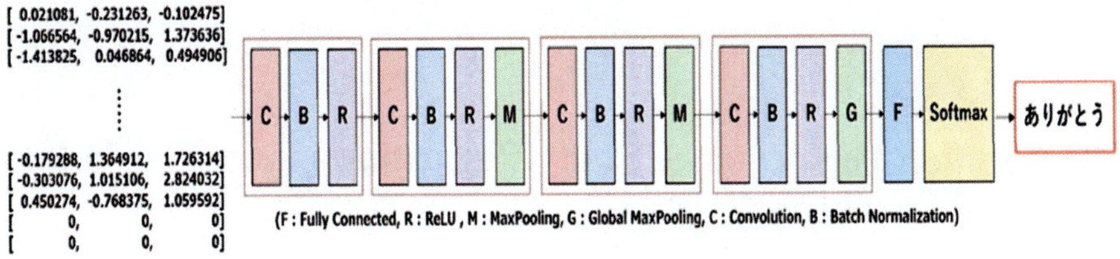

Fig. 5. Word recognition model

For example, in the case of word "A-RI-GA-TO-U", viseme sequence is "X, XA, A, AI, I, IA, A, AU, UO, O, OX, X". Then the VAE feature vector sequences were generated. Figure 5 shows the word recognition model used for our experiment. Normally, RNN is used for the time series data analysis, however, by using CNN and learning the changes of time series data, the network is able to perform like RNN.

We performed a simple 20word recognition test using our PC-based system. Using data augmentation techniques, the word recognition model was trained with 6000 datasets. Table 2 shows the word recognition results. Colored words were correctly recognized ones. From the test result, we confirmed the possibility of speaker dependent small vocabulary word recognition by simply using the VAE feature of 36 viseme sequences [16–18].

Table 2. 20 frequently used word recognition result (1st and 2nd candidate).

	True	Result	
		First	Second
1	A-RI-GA-TO	A-RI-GA-TO	O-HA-YO
2	I-I-E	I-I-E	KI-I-RO
3	O-HA-YO	O-HA-YO	A-O
4	O-ME-DE-TO	A-RI-GA-TO	O-ME-DE-TO
5	GO-ME-N-NA-SA-I	GO-ME-N-NA-SA-I	O-ME-DE-TO
6	SU-MI-MA-SE-N	SU-MI-MA-SE-N	I-I-E
7	DO-I-TA-SHI-MA-SHI-TE	DO-I-TA-SHI-MA-SHI-TE	GO-ME-N-NA-SA-I
8	HA-I	SA-N	HA-I
9	MO-SHI-MO-SHI	MO-SHI-MO-SHI	SA-N
10	A-KA	SA-N	A-KA
11	A-O	O-HA-YO	A-O
12	KI-I-RO	KI-I-RO	SI-RO
13	MI-DO-RI	MI-DO-RI	I-CHI
14	SHI-RO	KI-I-RO	SHI-RO
15	ZE-RO	ZE-RO	SA-N
16	I-TI	I-TI	NI
17	NI	I-TI	NI
18	SA-N	SA-N	HA-I
19	YO-N	GO	YO-N
20	GO	GO	YO-N

3 Recognition Performance Study Regarding Users, Vocabulary Size, and Speaking Style

We asked multiple subjects to perform a 20word recognition experiment without asking any specific speaking style, and compared their recognition accuracy. 36 viseme images and those 20-word utterance images were recorded. First, we investigated the recognition accuracy when no instruction was given. Table 3 shows the results of word recognition performed by four subjects. The word recognition accuracy varies greatly depending on the speaker. Then, seven subjects were asked to make the mouth shape clearly by paying attention to the difference in vowel mouth shape, and also they were asked to speak each syllable 50 times per minute. Table 4 shows the results of word recognition performed by those seven subjects. The second experimental result shows that subjects obtained around 60% accuracy (up to the sixth candidate).

Table 3. Word recognition result with no training

	Recognition result			
	h1	n1	o1	y4
First	4.4%	68.5%	3.8%	29.4%
Second	7.0%	78.2%	6.4%	41.0%
Third	14.1%	78.8%	8.3%	51.2%
Fourth	17.3%	82.0%	12.8%	64.1%
Fifth	23.0%	83.9%	18.5%	73.0%
Sixth	30.1%	86.5%	26.9%	79.4%

Table 4. Word recognition result after adjusting the mouth shape and the speech rate

	Recognition result						
	h2	n2	o2	y7	k1	s1	t1
First	17.3%	32.0%	25.6%	28.8%	26.2%	14.7%	21.1%
Second	27.5%	41.0%	39.7%	39.7%	43.5%	24.3%	33.3%
Third	35.8%	45.5%	52.5%	44.8%	54.4%	35.2%	40.3%
Fourth	46.1%	51.2%	57.0%	50.6%	58.3%	39.7%	43.5%
Fifth	51.9%	53.8%	61.5%	55.1%	60.2%	48.7%	48.0%
Sixth	60.2%	55.7%	64.7%	59.6%	65.3%	55.7%	51.2%

The vocabulary size was increased from 20 words to 40 words, and the recognition accuracy was tested. The experiment was conducted twice with one well trained subject. Table 5 shows the word recognition results. The recognition rate was 54% up to the third candidate and 65% up to the sixth candidate. The 20 words recognition test and the 40 words test results show the recognition performance difference was not so large. Those 40 words covers simple greetings and expressions that are often used in daily life. From those over all word recognition test results, the degradation due to increasing the vocabulary size is not so significant, however, further verification is required. As for the letter to viseme sequence conversion rules, we found that there is some mismatch between viseme sequence and actual mouth shape. That was the reason why "kyo" (today) was mis-recognized, for example. The word "kyo" was converted "X, XU, UO, O, O, OX, X". "kyo" was uttered "O", not "UO". Therefore, it is necessary to review the conversion rules especially when we increase the vocabulary size further.

Table 5. 40 Word recognition result.

	Recognition rate	
	yu1	yu2
First	31.4%	11.6%
Second	43.7%	22.5%
Third	54.1%	33.7%
Fourth	62.0%	41.6%
Fifth	64.1%	47.9%
Sixth	65.8%	51.2%

In summary;

(1) The user needs to aware what is the correct mouth shape, and the system should have some training mode to give users the training.
(2) A support function to be able to speak with a constant speed.
(3) It needs to tune the letter-to-viseme sequence conversion rule.

Moreover, since the recognition rate of the previous experiment was 95%, further analysis of the misrecognition factors is necessary.

4 Development of Lip-Reading System Using Mobile-Phone

In this study, we use Android Studio, Kotlin and Chaquopy Python SDK for Android. Camera2API was used for the camera application. The Camera2API allows you to preview images and capture video. As for the speech output, Text-to-Speech function in Android Studio was used. The target hardware is:

$$\text{Model}: \text{ SHARP SH - M0P}, \text{ CPU}: \text{SDM845}, \text{ RAM}: \text{ 4GB}$$

As for the specification, we are targeting the same or even better recognition accuracy as we obtained from the PC-based experimental system with 20 vocabulary words. We designed the user interface to be able to select multiple candidates easily. Figure 6 shows the screen image of the mobile device. We conducted the comparison test to see the recognition performance difference between those PC-based and mobile phone-based systems. We took the word set used in the study of Asami, in which LipNet was applied [13–16].

Table 6 shows the test result using the PC-based. At first, the recognition result using mobile phone system showed 95% accuracy including from the first to sixth candidate. That is almost similar to the PC-based system. However, the accuracy for the first candidate was 40%, which is much lower than the PC-based system. The main reason is that the frame rate difference is 1/3 smaller compare with the PC-based. Using

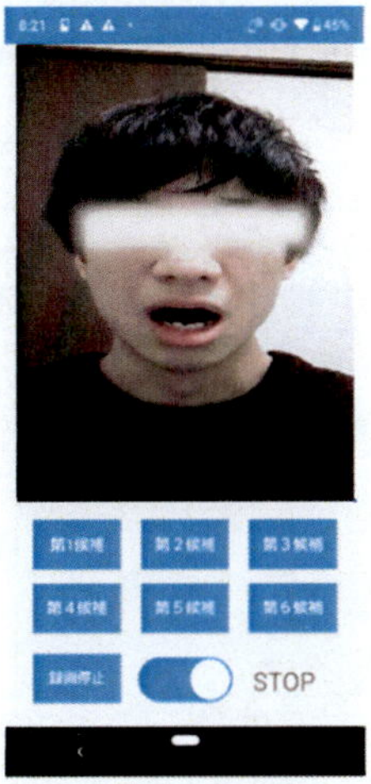

Fig. 6. Screen image of the mobile-phone system

interpolation method, we tested the same comparison study again and conformed the improvement as shown in Table 7.

As the result, the recognition accuracy was improved, but still not as good as the PC-based. The processing time is about 200 ms after start the recognition.

Table 6. 20 Word recognition test result using PC-based system.

	TRUE	Recognition result		
		First	Second	Third
1	A-RI-GA-TO-U	O-HA-YO-U	MO-SHI-MO-SHI	KI-I-RO
2	I-I-E	I-I-E	SHI-RO	ZE-RO
3	O-HA-YO-U	KI-I-RO	GO-ME-N-NA-SA-I	O-ME-DE-TO-U
4	O-ME-DE-TO-U	KI-I-RO	GO-ME-N-NA-SA-I	A-KA
5	GO-ME-N-NA-SA-I	GO-ME-N-NA-SA-I	KI-I-OR	MO-SHI-MO-SHI
6	SU-MI-MA-SE-N	SU-MI-MA-SE-N	O-ME-DE-TO-U	MO-SHI-MO-SHI
7	DO-U-I-TA-SHI-MA-SHI-TE	DO-U-I-TA-SHI-MA-SHI-TE	GO-ME-N-NA-SA-I	MO-SHI-MO-SHI
8	HA-I	HA-I	A-RI-GA-TO-U	I-CHI
9	MO-SHI-MO-SHI	MO-SHI-MO-SHI	MI-DO-RI	GO-ME-N-NA-SA-I
10	A-KA	A-KA	A-O	GO
11	A-O	A-O	GO	YO-N
12	KI-I-RO	KI-I-RO	HA-I	MO-SHI-MO-SHI
13	MI-DO-RI	MI-DO-RI	KI-I-OR	HA-I
14	SHI-RO	SHI-RO	ZE-RO	I-I-E
15	ZE-RO	SHI-RO	GO	YO-N
16	I-CHI	I-CHI	NI	HA-I
17	NI	NI	I-CHI	HA-I
18	SA-N	SA-N	A-KA	NI
19	YO-N	YO-N	GO	SA-N
20	GO	GO	A-KA	YO-N

Table 7. 20 word recognition test result using mobile phone application.

	TRUE	Recognition result		
		First	Second	Third
1	A-RI-GA-TO-U	A-KA	A-O	HA-I
2	I-I-E	ZE-RO	I-I-E	YO-N
3	O-HA-YO-U	KI-I-RO	MO-SHI-MO-SHI	GO-ME-N-NA-SA-I
4	O-ME-DE-TO-U	ZE-RO	YO-N	I-I-E
5	GO-ME-N-NA-SA-I	KI-I-RO	MO-SHI-MO-SHI	GO-ME-N-NA-SA-I
6	SU-MI-MA-SE-N	SU-MI-MA-SE-N	A-RI-GA-TO-U	HA-I
7	DO-U-I-TA-SHI-MA-SHI-TE	SU-MI-MA-SE-N	O-ME-DE-TO-U	MO-SHI-MO-SHI
8	HA-I	HA-I	A-RI-GA-TO-U	SU-MI-MA-SE-N
9	MO-SHI-MO-SHI	KI-I-RO	MO-SHI-MO-SHI	GO-ME-N-NA-SA-I
10	A-KA	A-KA	A-O	GO
11	A-O	A-O	A-KA	GO
12	KI-I-RO	KI-I-RO	GO-ME-N-NA-SA-I	MI-DO-RI
13	MI-DO-RI	ZE-RO	KI-I-RO	MI-DO-RI
14	SHI-RO	GO	NI	A-KA
15	ZE-RO	ZE-RO	YO-N	SHI-RO
16	I-CHI	NI	HA-I	I-CHI
17	NI	NI	I-CHI	GO
18	SA-N	GO	A-KA	NI
19	YO-N	GO	A-KA	A-O
20	GO	GO	A-KA	A-O

5 Discussion

Currently, we have been developing word/phrase level lip-reading system for laryngectomees to be able to communicate more easily. However, our 40 words recognition result shows that the accuracy is still around 60% even if we look at the candidates up to 6th place. Only well trained users could reach around 95%. In this study, we also implemented and tested the mobile-phone-based lip-reading system. The recognition accuracy seems almost equivalent to the PC-system, and the response speed is reasonably fast (200 ms). That is encouraging.

There are a couple of possible improvement method we plan to work on. Firstly, considering the current algorithm is speaker-dependent, vocabulary independent system, it is possible to generate any word/phrase reference pattern by concatenating the 36 veseme feature vectors. Therefore user can switch the word reference dictionary depend on the scene. Secondly, a simple training application using mobile-phone can be effective if you see the experimental result between the Table 3 and Table 4. By reforming the mouth shape and stabilize the speech rate shows significant improvement. Thirdly, next word/phrase prediction algorithm can be implemented using the wordlist buttons on the screen.

As for the basic research, we would like to find an innovative way to recognize consonants from the lip region images.

6 Conclusion

In this study, we performed a couple of lip-reading-based word recognition performance experiments regarding multiple users, vocabulary size, and speaking style for laryngectomees. By reforming the mouth shape and stabilize the speech rate showed significant improvement. Also, a mobile-phone-based prototype and the user interface were developed and tested the effectiveness. As for the lip-reading algorithm, viseme sequence representation with VAE were used to be able to adapt users with very small amount of training data set. In the case of the mobile-phone-based lip-reading system, the recognition accuracy seems almost equivalent to the PC-system, and the response speed is reasonably fast (200 ms). The experimental result showed around 60% recognition accuracy including candidates up to six place under the condition of 20word vocabulary size with seven subjects. In case of 20 word and 40word vocabulary size comparison, one well trained subject obtained almost the same result. For future study, considering the current algorithm is speaker-dependent, vocabulary independent system, we would like to test switching the word reference dictionary depend on the scene. Also, a simple training application using mobile-phone can be effective. We plan to design effective training method by reforming the mouth shape and stabilize the speech rate. As for the basic research, we would like to find a innovative way to recognize consonants from the lip region images.

Acknowledgments. This study was subsidized by JSPS Grant-in-Aid for Scientific Research 1 9K12905.

References

1. Matsui, K., et al.: Enhancement of esophageal speech using formant synthesis. J. Acoust. Soc. Jpn. (E) **23**(2), 66–79 (2002)
2. Matsui, K., et al.: Development of speech enhancement system. IEEJ J. **134**(4), 216–219 (2014)
3. Kimura, K., et al.: Development of wearable speech enhancement system for laryngectomees. In: NCSP 2016, pp. 339–342, March (2016)
4. Nakahara, T., et al.: Speech enhancement system using lip-reading. In: 17th International Conference on Distributed Computing and Artificial Intelligence, DCAI 2020, pp. 159–167, October (2020)
5. Matsui, K., et al.: Mobile device-based speech enhancement system using lip-reading. In: IICAIET 2020. September (2020)
6. Denby, B., Schultz, T., Honda, K., Hueber, T., Gilbert, J.M., et al.: Silent speech interfaces. Speech Commun. **52**(4), 270 (2010)
7. Kapur, A., Kapur, S., Maes, P.: AlterEgo: a personalized wearable silent speech interface. In: IUI 2018. Tokyo, Japan, March 7–11 (2018)
8. Goodfellow, I., Bengio, Y., Courville, A.: Deep Leaning. MIT Press, Cambridge, Massachusetts (2016)
9. Saito, Y.: Deep Learning From Scratch. O'Reilly, Japan (2016)
10. Hideki, A., et al.: Deep Leaning. Kindai Kagakusya, Tokyo (2015)
11. King, D.E.: Max-Margin Object Detection. arXiv:1502.00046v1 [cs.CV], 31, Jan (2015)
12. Kazemi, V., Sullivan, J.: One millisecond face alignment with an ensemble of regression trees, In: IEEE Conference on Computer Vision and Pattern Recognition, pp. 1867–1874 (2014)
13. Assael, Y.M., Shillingford, B., Whiteson, S., de Freitas, N.: LipNet: end-to-end sentence-level lip reading. In: GPU Technology Conference (2017)
14. Ito, D., Takiguchi, T., Ariki, Y.: Lip image to speech conversion using LipNet. Acoustic Society of Japan articles, March (2018)

15. Kawahara, H.: STRAIGHT, exploitation of the other aspect of vocoder: perceptually isomorphic decomposition of speech sounds. Acoust. Sci. Technol. **27**(6), 349–353 (2006)
16. Asami, et al.: Basic study on lip reading for Japanese speaker by machine learning. In: 33rd, Picture Coding Symposium (PCSJ/IMPS2018), pp. 3–8. Nov. (2018)
17. Saitoh, T., Kubokawa, M.: SSSD: Japanese speech scene database by smart device for visual speech recognition. IEICE **117**(513), 163–168 (2018)
18. Saitoh, T., Kubokawa, M.: SSSD: speech scene database by smart device for visual speech recognition. In: Proceeding of the ICPR 2018, pp. 3228–3232 (2018)

Author Index

© The Editor(s) (if applicable) and The Author(s), under exclusive license
to Springer Nature Switzerland AG 2022

K. Matsui et al. (Eds.): DCAI 2021, LNNS 327, pp. 221–222, 2022.
https://doi.org/10.1007/978-3-030-86261-9